Strategic Safety Management in Construction and Engineering

Strategic Safety Management in Construction and Engineering

Patrick X.W. Zou
Professor of Construction Engineering and Management,
Swinburne University of Technology, Australia

Riza Yosia Sunindijo
Lecturer in Construction Management and Property
Faculty of the Built Environment
UNSW Australia (The University of New South Wales)

WILEY Blackwell

Library of Congress Cataloging-in-Publication Data applied for

A catalogue record for this book is available from the British Library.

ISBN: 9781118839379

Set in 10/12pt, MinionPro by Laserwords Private Limited, Chennai, India

1   2015

Contents

Foreword

Two thousand years ago, the Roman Statesman Marcus Tullius Cicero argued that "the safety of the people shall be the highest law." An emphasis on health and safety and the protection of all human beings should be the mark of any advanced society. However, somehow, health and safety (or "elf'n safety") seems to have acquired a negative, restricting reputation. Most of my new undergraduate students admit that they have had good, exciting activities in their schools and colleges cancelled because of "elf'n safety". And yet, in a review of the exemplary success on all fronts of the construction of the London 2012 Olympic Park, General the Lord Dannatt (*British Army Chief of the General Staff 2006–2009*) found that "health and safety was not just an annoying millstone hung around middle management's neck, but it was the enabling theme on which the project senior leadership team could found the bedrock of operational efficiency leading to completion under budget and ahead of schedule."

This book presents a strategic perspective on construction safety management providing both a historical and contemporary commentary. It deals with economics, climate and culture, skills, training and learning as well as the important contemporary topic of safety in design. The book also explores research methods in the domain and the research to practice challenge.

I have known the author for many years and been privileged to learn more about his work in this important area.

I commend this book to you.

Alistair Gibb, PhD, BSc, CEng, MICE, MCIOB
European Construction Institute,
Royal Academy of Engineering Professor
Loughborough University, UK

Acknowledgements

Writing a book such as this means we have drawn data from a large number of sources, and we are indebted to many experts and commentators for their help. Especially we would like to thank Emeritus Professor Denny McGeorge of the University of Newcastle for proofreading the chapters; Adam Sun, a former MPhil student at UNSW Australia, for relevant data collection and analysis efforts for Chapter 2; Professor Andrew Dainty of Loughborough University for his invaluable contributions to Chapter 7; and Professor Alistair Gibb of Loughborough University for writing the Foreword.

We are grateful for the information obtained through various sources from the organisations included in this book: Lend Lease, Fluor, Gammon, John Holland, the Master Builders Association of the Australia Capital Territory, Leighton Holdings and so on.

The authors would like to thank UNSW Australia, University of Canberra and Swinburne University of Technology, where they have worked for many years, including undertaking part of the safety research that has been incorporated in this book.

We are also thankful to the editor(s) at Wiley–Blackwell for their contributions, especially Dr Paul Sayer and Harriet Konishi.

Finally, we would like to dedicate this book to our parents and family who have supported us continuously.

1 Safety Management in Construction and Engineering: An Introduction

This book addresses *Safety Management in Construction and Engineering* by taking a broad view of safety from a strategic decision-making and management perspective. It focuses on strategic decisions made by the boardroom and senior management, including safety strategy design, development, implementation and evaluation. The book also addresses the importance of balancing and integrating the 'science' and 'art' of safety management, together with an exploration of how safety is perceived and enacted by top management and on-site operatives. The localised on-project-site context for safety strategy implementation, monitoring and evaluation is emphasised, while case studies are provided to demonstrate the implementation of safety concepts, principles and techniques in practice.

The importance of the industry

Construction and engineering is an US $8.7 trillion market, accounting for 12.2% of the world's economic output (Global Construction Perspectives & Oxford Economics, 2013) and providing employment for about 200 million people worldwide (Murie, 2007). It is supported by a complex supply chain encompassing numerous industries ranging from steel, timber and concrete producers to furniture and carpet manufacturers. The supply chain extends further to other industries, such as trucking, shipping, manufacturing and mining, which may not have an obvious direct relationship to the construction and engineering industry (Hampson & Brandon, 2004; Jackson, 2010). The industry is important because of its size and output, which underpins various

Strategic Safety Management in Construction and Engineering, First Edition.
Patrick X.W. Zou and Riza Yosia Sunindijo.
© 2015 John Wiley & Sons, Ltd. Published 2015 by John Wiley & Sons, Ltd.

economic activities and contributes to the delivery of social and environmental objectives of a nation (Health and Safety Executive, 2009).

By way of demonstrating the importance of the construction and engineering sector, Australia and the UK are cited as examples. In Australia, the construction and engineering industry engages in three broad areas of activity: residential building (houses, apartments, flats, and so on), non-residential building (offices, shops, hotels, schools, and so on), and engineering construction (roads, bridges, rails, water and sewerage, and so on). Both the private and public sectors undertake construction and engineering activities. The private sector is engaged in all three categories, while the public sector plays a key role in initiating and undertaking engineering construction activities and those related to health and education (ABS, 2010). The construction and engineering industry is the third largest contributor to Gross Domestic Product (GDP) in the Australian economy and has a major role in determining economic growth. In 2010–11, the industry accounted for 7.7% of GDP and had significantly increased its share of GDP from 6.2% in 2002–03. It also employed 9.1% of the Australian workforce in 2010–11, making it Australia's third largest industry after health care and social assistance, and retail trade (ABS, 2012).

In the UK, the construction and engineering industry contributes about 6.7% to the nation's economy and 10% of all jobs. The UK also has the sixth largest green construction sector in the world. Due to the importance of the sector, the UK government published the Construction 2025 report, which summarises the industrial strategy for the construction sector in the coming decade. The Construction 2025 report outlines the steps that the government and the industry will take in the short and medium terms to achieve four ambitious goals: (1) a 33% reduction in both the initial cost of construction and the whole life cost of assets, based on 2009–10 levels, (2) a 50% reduction in the overall time from inception to completion for new buildings and refurbished assets, based on the industry's performance in 2013, (3) a 50% reduction in greenhouse gas emissions in the built environment as compared to the 1990 baseline, and (4) a 50% reduction in the trade gap between total exports and total imports for construction products and materials based on data in February 2013. The UK government also stresses the importance of investment in infrastructure projects and house building for the economy (HM Government, 2013).

Characteristics of the construction and engineering sector

The construction and engineering sector has unique characteristics which influence the ways construction and engineering organisations operate within the sector, including how they manage safety. These characteristics can be classified into two levels: industry-related and project-related, as discussed in the following sections.

Industry-related characteristics

There are several characteristics at this level, which influence an organisation as a whole. They typically reflect the conditions and nature of the industry. The first characteristic is that the industry is complex in nature. In 1996, Gidado (1996) explained (and it is still valid today) that this complexity originates from (1) uncertainty due to the various components needed in each activity within the production process, which come from various sources including the resources employed and the environment, and (2) interdependence among activities, which is concerned with bringing different parts together to form a work flow. Gidado (1996) further elaborated that the uncertainty has four causes: (1) the unfamiliarity of management with local resources and the local environment, (2) lack of complete specification for the activities on site, (3) the uniqueness of every construction and engineering project (with regard to materials used, type of work, project teams, location and time) and (4) the unpredictability of the environment. This uncertainty characteristic compels construction and engineering organisations to apply a decentralised approach to decision making. The interdependence is influenced by three factors: (1) the number of technologies and their interdependence, (2) the rigidity of sequence between various main operations and (3) the overlaps of stages or elements in construction and engineering processes. The organisation of the workforce into trades and the subcontracting practice intensify this interdependence, which calls for more local rather than centralised coordination (Dubois & Gadde, 2002).

The second characteristic is the low levels of entry to and exit from the construction industry and the large number of small-size enterprises. Although considered a service industry, the entry to the construction industry is different from other service industries, such as finance, insurance, real estate, professional services and business services. The level of education is the most important factor in identifying entrants into self-employment in the other service sectors, but it is less so for construction. In fact, high school dropouts are much more likely to enter self-employment in construction than college graduates (Bates, 1995). In Australia, about 90% of construction organisations employ less than five people or are identified themselves as sole proprietorships. In 2013, the construction industry had the highest number of businesses operating in Australia. Within the same period, however, there were also more than 50,000 exits, representing a 16.5% exit rate, which is higher than the average for all industries at 14.1% (ABS, 2013). All this indicates that there are low requirements to enter the industry, while at the same time the exit rate is also relatively high, thus demonstrating the dynamic nature of the industry.

The third characteristic is the intense and fierce competition and low profit margins due to the sheer number of construction and engineering businesses, especially the small-sized ones (Arditi et al., 2000).

The fourth characteristic is economic pressures, which are typically worsened by late progress payments, and unfair allocations of risk (Arditi et al., 2000;

Duffy & Duffy, 2014), which lead to confrontational relationships between parties, making the industry well known for its reputation for fragmentation, conflicts, mistrust, claims and litigation (Duffy & Duffy, 2014; Kanji & Wong, 1998).

The fifth characteristic is related to the workforce, which is labour intensive. Despite the effort to automate and the general advancement of technology, the industry remains traditional and is slow in adopting new technology. Many construction sites still use relatively high rates of unskilled workers, especially those in developing countries (Giang & Pheng, 2011). Furthermore, recent trends show an increase in the proportion of older workers. Together with the physically demanding nature of the construction and engineering work and the exposure to the external environment when working on sites, they intensify the already challenging job and increase the risk of injury and chronic health conditions among older workers (Brenner & Ahern, 2000; Schwatka et al., 2012).

The sixth characteristic is gender imbalance. The industry is one of the most gender-segregated sectors where men dominate the employment in the building trades. The sector is considered as 'tough' due to this masculine identity. The culture of taking safety risks and working physically for long hours in primitive working conditions are considered as the norms. This 'man's job' is associated with physical labour, dirt, discomfort and danger, which, interestingly, creates a hierarchy within the building trades. The rougher, dirtier trades are perceived to be more masculine than the more 'refined' and intellectual trades. Labourers, steel-fixers, bricklayers and ground workers are at the bottom of the status hierarchy, but at the top of the masculinity hierarchy. In contrast, electricians have a high status, but they are not real men (Ness, 2012). This kind of mindset spawns resistance to change, making it extremely difficult to persuade the construction workforce to embrace safety, which is considered as an intrusion into their 'normal' ways of operating (Lingard & Rowlinson, 2005).

Project-related characteristics

The construction and engineering industry is also inherently a site-specific project-based activity. According to Project Management Institute (2013), a project has two characteristics: temporary and unique. First, temporary indicates that a project has a definite beginning and end in nature. The end is reached when project objectives have been achieved or when the project is terminated because its objectives cannot be met, or when the need for the project no longer exists. A project may also be terminated if the client wishes to do so.

Second, every project creates a unique result. Buildings can be constructed with the same or similar materials, by the same or different teams and by the same or different methods of construction. There are many factors that cause each construction project to be unique, such as the site location, design, specific circumstances, stakeholders and so on. These characteristics provide added challenges in construction organisations and the industry at large. While many other industries have standardised their elements and activities, the construction sector has been slow in adopting standardisation. The uniqueness of each project and project constraints may also make standardisation difficult (Dubois & Gadde, 2002; Kanji & Wong, 1998). Furthermore, the lack of standardisation, the unique nature of each project and the temporary nature of a project limit the impact of learning because project teams need to re-learn and contextualise their learning every time they move to a new project.

The third characteristic is related to the uncertain and interdependent environment of the construction and engineering industry, as discussed in the previous section, which causes more problems when it comes to achieving project objectives. In this kind of environment, failure of any of the parties may seriously affect project duration and the quality of the final product. The traditional project delivery system, which separates the design and construction stages, may result in the lack of constructability and excessive design changes during construction, which escalate costs and delay the project (Arditi et al., 2000; Kanji & Wong, 1998).

Fourth, the strong emphasis on individual projects favours a narrow perspective, both in time and scope (Dubois & Gadde, 2002). As a result, competitive tendering is seen as a strategy to promote efficiency and to assure that a job is carried out at the lowest possible cost. Instead of generating efficiency, this practice can lead to poor performance. Contractors may cut corners to reduce their tender prices only to use variations due to incomplete design or project changes to inflate the prices in the later stage of construction. Awarding contracts based on price alone may also lead to reduced quality, conflicts, and poor safety implementation. Mayhew and Quinlan (1997) indicated that competitive tendering worsens project safety risks because economic pressures and intense competition penalise those organisations who try to do the right thing due to their higher tender prices. Although the industry may have slowly shifted to the value-for-money basis in assessing project tenders, generally price is still the main factor that decides the winning bid.

Fifth, mega projects typically have significant impacts on the surrounding community and environment. McDonald-Wilmsen (2009) estimated that 90 million people might have been displaced in the past decade due to such projects. For example, the development of the Three Gorges Dam project required the Chinese government to resettle more than one million people, which led to many socio-cultural issues, such as ineffective compensation distribution, loss of employment and inadequate new housing (McDonald-Wilmsen, 2009). This project also has significant on-going impacts on the environment. The increased water levels in the Three Gorges Dam (up to 175 metres) may destabilise the soil and cause landslides as well as other environmental impacts, which could be hazardous to aquatic flora and fauna. Eventually, species that cannot adapt to the new environment will disappear (Stone, 2008).

There are also other features affecting safety in construction projects, such as design complexity, tight project duration, multiple level subcontracting and construction, site restriction and complicated procurement and contracting systems.

Why a book on strategic safety management?

Today's major construction organisations recognise the need to integrate safety into all decision making. We believe that strategic safety management is a way of achieving the level of integration which is indicative of a mature organisation. Many developed economies have made significant improvement in safety management through the use of systems, structures and modern technology, but have found it difficult to achieve exponential improvements in safety performance. More of the same will not produce the next big leap in safety performance (Wagner, 2010). This is because no matter how automated a production process or complex a management system is, people cannot be separated from the process or the system. People still control production and sometimes must intervene when unplanned events occur. It is often concluded that human error is the cause of 50–90% of all accidents. Simply put, people make mistakes and while human error may be undesirable, it is an inevitable aspect of everyday life (Lingard & Rowlinson, 2005; Peters & Peters, 2006). Reason (1990) argued that safety improvement can only be achieved through an attention to human error mechanisms. This human factor is particularly important in the construction and engineering industry due to its labour-intensive characteristic (Lingard & Rowlinson, 2005). Consequently, construction and engineering organisations should recognise the need to balance and integrate 'science' and 'art' when implementing safety management.

The premise of this book is that viewing safety management from the strategic perspective serves as a unique view point that has not been covered in the current body of safety knowledge. The following topics, portraying different components of the 'science' and 'art' of safety management, will be discussed in the subsequent chapters, demonstrating how these key components can be integrated into construction and engineering businesses and projects. Case studies and examples are also given in each chapter to show how each topic can be applied in practice. Figure 1.1 is a conceptual model which illustrates the relationship of strategic safety management to the other themes contained in this book.

Historical development and current trends in construction safety management

Construction safety has undergone a substantial evolution in the past century as illustrated in Figure 1.2. In the early 1900s, safety was virtually non-existent. There were no workers' compensation laws, thus typically construction organisations did not need to pay anything when accidents happened. Without any compelling financial incentive, there was no encouragement for the industry to implement or consider safety. It is difficult to forget the iconic black and white photograph, taken in 1932, showing workers sitting on a steel beam without any personal protective equipment (PPE) 69 floors up in the Rockefeller

Figure 1.1   Key themes in this book

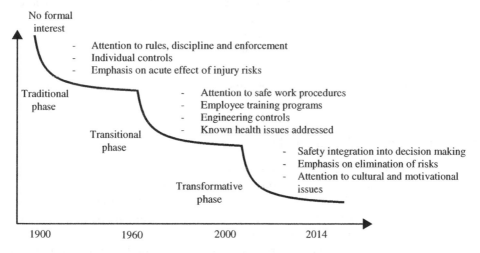

Figure 1.2   The evolution of safety management   (Source: adapted from Pybus, 1996, p.18 in Finneran and Gibb, 2013.)

Center project (Rockefeller Center, 2013). About the same time, workers scaled the structure of the Sydney Harbour Bridge without any fall protection (NSW Government, 2010). An infamous tragic accident showing the lack of interest towards safety happened in 1911 in New York when a fire broke out in the Triangle Waist Company building killing 146 employees, mostly women. It is believed that the exit doors were deliberately locked, the fire escape was dysfunctional, and the fire-fighting equipment was insufficient, further demonstrating the lack of safety concerns during this period (Cornell University, 2004).

Protesting voices arose, bewildered and angry at the lack of concern and the greed that caused these catastrophes. Responding to this demand, workers' compensation law was passed, thus compelling many industrial sectors to improve their safety performance. This became the beginning of the traditional phase in safety management. In the early years of the safety movement, the management focused on improving physical conditions and reducing unsafe behaviours. Workers were required to follow a set of rules and to use personal protective equipment at work (Petersen, 1988).

After this initial step, safety professionals started thinking in management terms, marking the dawn of the transitional phase. Initially, setting policies, defining responsibilities and clarifying authorities became a trend. In the 1960s and 1970s, professionalism was the focus and it was achieved by defining the scope and functions of safety professionals, developing curriculums for formal safety education and establishing a professional certification program. Some governments also passed more stringent occupational health and safety acts, further compelling the industry to take safety measures seriously. Many organisations became more proactive in implementing safety by establishing safety plans, safe work procedures and identifying safety risks before any activity began (Lingard & Rowlinson, 2005; Petersen, 1988).

The current trend is at the transformative phase, which places emphasis on the integration of safety into decision making, the elimination of safety risks, the development of workforce skills and the fostering of safety culture as we described in the previous section, demonstrating the importance of strategic safety management.

Alarming incident and injury problems

This evolution of safety has significantly improved safety performance in the construction industry. However, in recent years, it appears that this improvement has plateaued and the industry is facing difficulties in achieving further improvements, while injuries and fatalities still occur on a regular basis. Despite having an important role in the global and national economies, the construction industry has a notorious reputation as being one of the most dangerous industrial sectors (Health and Safety Executive, 2013; Lingard & Rowlinson, 2005; Murie, 2007; Safe Work Australia, 2013). It provides employment for about 7% of the world's workforce, but is responsible for 30–40% of work-based fatal injuries (Murie, 2007). The International Labour Organisation (2003) estimated that there are at least 60,000 fatalities on building sites every year. This estimate is conservative because many countries underreport their construction injuries and fatalities. Murie (2007) estimated that 100,000 people are killed on construction sites annually.

In the UK, the statistics for 2012–13 showed that the construction industry accounts for only about 5% of employment, but is responsible for 27% of fatal injuries and 10% of reported major injuries. The rate of fatal injury per 100,000 workers was 1.9 and the industry still accounts for the greatest number of fatal injuries among the industrial sectors (Health and Safety Executive, 2013). In Australia, there were 211 fatalities in the construction industry from 2007 to 12, corresponding to 4.34 fatalities per 100,000 workers, which is nearly twice the average national fatalities rate of 2.29 (Safe Work Australia, 2013). In Singapore, the fatality rate in 2013 was 2.1 per 100,000 workers, while the rate in the construction industry was more than three times higher at 7.0 (Ministry of Manpower, 2013). These statistics are worse in the USA where, in 2012, the rate of fatal injury per 100,000 workers in the construction industry was 9.9, significantly higher than the average rate for all industries at 3.4 and the rate in other developed nations (Bureau of Labor Statistics, 2014).

The above-mentioned statistics and the six reasons listed below (Holt, 2005; Reese & Eidson, 2006), highlight the need to continue improving safety performance in construction.

1. Governments around the world have laws that require construction organisations to provide safe work conditions and adequate supervision. Lack of safety, therefore, may lead to prosecution or claims, which will become the source of extra costs and adverse publicity.
2. Lack of safety increases the probability of accidents, which may lead to human suffering, disabilities and deaths.
3. When an accident happens, the morale of workers is weakened. On the contrary, accident prevention programs strengthen morale and improve on-site productivity.
4. A safe operation in the workplace is considered a moral obligation by the current society; thus good safety practices are essential to improve and maintain reputation.
5. A good safety record and proven safety management system are valuable marketing tools to attract new clients and support business expansion.
6. A safety management program contributes to the financial health of construction organisations by helping them avoid costs associated with accidents. An accident incurs both direct and indirect costs as well as insured and uninsured costs.

Current safety management body of knowledge

Due to the high rates of accidents and injuries in the construction industry, much effort has been taken through research and better work practices to improve safety performance. Figure 1.3 shows the domains of research and practice aimed at promoting safety performance improvement in the industry. The overlaps between the circles and the ellipse illustrate the level of interface between each of these domains and the effort required to improve safety

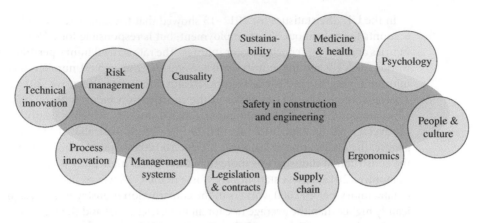

Figure 1.3   Indicative interfaces between safety and other domains of research and practice (Source: adapted from Finneran and Gibb, 2013.)

(Finneran & Gibb, 2013). For example, there is a significant overlap between the domains of risk and safety because the principles of risk management are commonly used to identify, assess and mitigate safety risks. On the other hand, there is a much less overlap between technical innovation and safety because of the traditional and labour-intensive characteristics of the industry. Recently, research on building information modelling (BIM) and its application in practice has grown in popularity.

There are many books on safety management, further demonstrating the importance of this topic. Some of these books, particularly the ones related to the construction and engineering industry, totalling more than 10 titles, are listed in Bibliography.

The book's contents

Chapter 2: economics of safety

This chapter discusses the economic aspects of safety management in construction, including benefits, costs and investment optimisation. This topic is significant because efforts to improve safety are hindered by many barriers and those influenced by economic considerations are particularly dominant. As stated earlier, the subcontracting practice in the industry, coupled with intense competition and the uncertainty of demand, force construction organisations to focus on reducing costs at the expense of other factors, including safety. Stakeholders in construction and engineering should realise that lack of safety increases the probability of accident occurrences and an accident could have an adverse impact on economic performance due to fine and compensation costs, loss of productivity, production delay, weakened morale and bad reputation.

On the contrary, safety investment and management have the potential to generate economic advantages for construction organisations by avoiding the costs related to accidents.

The role of clients, who have the economic power, is important to facilitate safety implementation. Without their support, construction and engineering organisations will face numerous constraints in implementing safety measures, because of the confrontational nature of the industry. Safety should become one of the selection criteria in the tender evaluation process and there should be mutual commitment to provide necessary resources to apply innovative safety measures when the opportunity arises. Through this effort, clients will also be able to enjoy the economic benefits of safety implementation. Despite the reality of the economics of safety, we believe that the fundamentals of safety are about the preservation of human life and the protection of the human right for a safe work environment. Although it may be possible to measure the economic benefits of safety, it is never appropriate to measure human life in monetary terms. Therefore, strategic safety management decisions should be taken from a moral, ethical and human right standpoint.

Chapter 3: safety climate and culture

This chapter discusses key concepts related to safety culture, including safety climate, safety culture dimensions, safety subcultures and safety culture maturity and its measurement. It also presents several case studies to demonstrate the best international practice in implementing safety programmes for fostering a strong safety culture in construction businesses and projects. Since its conception in 1986 after the Chernobyl disaster, safety culture has increased in popularity and its poor development has been constantly highlighted as the key source of major accidents. Over the years, the definition of safety culture has converged and the concept has become clearer with a solid theoretical underpinning. Fernández-Muñiz et al. (2007) offered a lengthy, yet comprehensive definition of safety culture, which is 'a set of values, perceptions, attitudes and patterns of behaviour with regard to safety shared by members of the organisation; as well as a set of policies, practices and procedures relating to the reduction of employees' exposure to occupational risks, implemented at every level of the organisation, and reflecting a high level of concern and commitment to the prevention of accidents and illnesses'.

Safety culture has three dimensions: psychological, behavioural and corporate. The psychological dimension refers to safety climate, which encompasses the attitudes and perceptions of employees towards safety and safety management systems. The behavioural dimension is concerned with what employees do within the organisation. The corporate dimension refers to the organisation's safety policies, operating procedures, management systems, control systems, communication flows and workflow systems (Health and Safety Executive, 2005). Developing a strong safety culture requires managers to focus on developing five sub-cultures: informed (collective alertness towards things that could go wrong), reporting (the readiness and willingness to report safety issues), just (the organisation's willingness to be accountable towards safety), learning (the willingness to learn and change), and flexible cultures (the ability to decentralise during emergency) (Hopkins, 2005; Reason, 2000). The safety culture maturity of an organisation can be measured by how the subcultures are manifested in each safety culture dimension. The future trends of safety culture are also included in the chapter.

Chapter 4: skills for safety

Although safety culture should be initiated by top-level managers, its development and implementation require the support of all the employees. Those in safety-critical positions should provide safety leadership to ensure that safety implementation is aligned from the top to the project and work site levels. This chapter discusses four management skills, comprising conceptual, human, political and technical skills, which are needed by project management personnel to perform their safety leadership role effectively. The discussion also includes 15 skill components that form the four sets of skills.

Our investigation found that visioning, self-awareness and apparent sincerity are foundational skill components. These are followed by the first-tier mediators, which consist of scoping and integration, and self-management. The second-tier mediators are relationship management, social awareness and social astuteness. Although technical skills are absent from the model, this does not mean that technical skill is unnecessary, but it becomes less important at the higher management levels where emphasis is on supervising and coordinating the work of others. The nature of the construction and engineering industry with its various stakeholders and elements also compels industry practitioners, especially those at management levels, to use more conceptual, human and political skills rather than technical skills. The focus of skill development, therefore, should be on these three 'soft' skills which are then supported by sufficient technical proficiency.

Chapter 5: safety training and learning

Safety training refers to programs and processes imposed externally on employees by regulatory bodies, the industry and organisations, whereas safety learning is concerned with the learning experience and learning processes of learners in training programs, at work sites, and via self-learning opportunities. Following from the previous topic, employees should be equipped with safety skills and knowledge which enable them to work safely and to encourage others to do the same. Safety training and learning equip employees with these skills and knowledge. The construction and engineering sector is characterised by temporary organisations and extensive outsourcing. Employees and workers move from one project to another more frequently than in other industrial sectors.

It is important for employees and workers to attain and retain safety skills and knowledge in order to improve safety performance in this dynamic industry.

This chapter discusses different approaches that construction organisations can use to advance a climate that values safety learning. Many organisations associate the learning process with pedagogical methods, such as lectures and presentations, assigned readings and examinations. This approach, although necessary, has been proved not to be particularly well-suited for adult learners, particularly those who have substantial work experience. The andragogy approach, which assumes that learners are self-directed and problem-centred in their learning, should be applied, together with existing pedagogical practice, to improve the effectiveness of safety training programs. We need to realise that besides learning from formal channels, safety learning is about taking part in the social world, that is, learning takes place among and through others at work (Gherardi & Nicolini, 2002). This chapter ends with a four-tiered technique to evaluate the effectiveness of safety training programmes.

Chapter 6: safety in design, risk management and BIM

Many studies have revealed that considering safety in the design stage, including architectural and engineering designs, has a great potential to significantly reduce the number of accidents during the construction stage. The separation between the designers and contractors in the construction and engineering industry is one of the key challenges in undertaking safety assessment at design. When designers have a lack of knowledge and experience in construction processes and materials, this separation hinders communication and may increase safety risks in construction and operational stages. This chapter discusses the theory and practice of safety in design, including the process, barriers and success factors, as well as legal and policy requirements. It also discusses knowledge and skills required for designers to undertake safety assessment at the design stage. Tools to facilitate implementation of the safety-in-design concept are also discussed. These tools include safety risk management and BIM. Case studies are also used to demonstrate the application of the safety-in-design concept in practice.

Chapter 7: safety research methods and research-practice nexus

Performing safety research is a viable strategy to ensure continuous safety performance improvement. This chapter presents three research methodologies (quantitative, qualitative and mixed methods) commonly used in social science research and the philosophical assumptions behind them. The analysis on recent safety-related research found that quantitative methodology was the dominant methodology. Recognising a principle established earlier (safety training and learning) in which safety learning also occurs in practice through interactions with people and artefacts at work, we encourage the use of qualitative research to develop an in-depth understanding of strategic safety management in practical settings. Furthermore, although much safety research has been undertaken, there is a danger that research results are not communicated to workers or, in the worst case scenario, are not relevant to practice. The integration of the realms of theory and practice is crucial to ensure the practicality of research findings, which leads to real safety improvement. Consequently this chapter proposes a mixed-methods research design to exploit the strengths of both quantitative and qualitative methodologies, and to achieve a safety-research practice nexus through iteration between the realms of theory and practice to promote co-production of knowledge by researchers and practitioners.

Chapter 8: strategic safety management

This chapter consolidates all the topics into a strategic safety management framework, which consists of two interrelated dimensions: the 'science' and 'art' of safety management. This chapter also discusses the process of strategic safety management, which includes strategy development, strategy implementation and strategy evaluation. Developing safety strategies starts from the foundation, which is to establish safety vision, goals and core competencies. Based on this foundation, the contents of strategic safety management, that is, all the elements discussed in the previous chapters, can be developed and contextualised according to each organisation's unique condition. Implementing safety strategies is about having corporate governance to ensure that strategic decisions are made effectively, an appropriate organisational structure so that strategy implementation is aligned across management levels and strategic leadership to develop safety commitment in all employees. A unique approach proposed to evaluate safety strategies is to use the balanced scorecard method to link an organisation's long-term strategies with its short-term actions. A detailed case study is provided to demonstrate how a strategic safety management process can be applied in practice.

References

ABS. (2010). *Topics @ a Glance – Construction*. Belconnen: Australian Bureau of Statistics.

ABS. (2012). *1301.0 – Year Book Australia, 2012*. Belconnen: Australian Bureau of Statistics.

ABS. (2013). *Counts of Australian Business, including Entries and Exits*. Canberra: Australian Bureau of Statistics.

Arditi, D., Koksal, A., & Kale, S. (2000). Business failure in the construction industry. *Engineering, Construction and Architectural Management, 7*(2), 120–132.

Bates, T. (1995). Self-employment entry across industry groups. *Journal of Business Venturing, 10*(2), 143–156.

Brenner, H., & Ahern, W. (2000). Sickness absence and early retirement on health grounds in the construction industry in Ireland. *Occupational & Environmental Medicine, 57*(6), 615–620.

Bureau of Labor Statistics. (2014). Census of Fatal Occupational Injuries (CFOI) – Current and Revised Data Retrieved 23 July 2014, from http://www.bls.gov/iif/oshcfoi1.htm

Cornell University. (2004). The Triangle Factory Fire Retrieved 18 December 2009, from http://www.ilr.cornell.edu/trianglefire/

Dubois, A., & Gadde, L.-E. (2002). The construction industry as a loosely coupled system: Implications for productivity and innovation. *Construction Management and Economics, 20*(7), 621–631.

Duffy, S., & Duffy, J. (2014). An analysis of dispute review boards and settlement mediation as used in the Australian construction industry. *Building and Construction Law Journal, 30*(3), 165–174.

Fernández-Muñiz, B., Montes-Peón, J. M., & Vázquez-Ordás, C. J. (2007). Safety culture: Analysis of the causal relationships between its key dimensions. *Journal of Safety Research, 38*(6), 627–641.

Finneran, A., & Gibb, A. (2013). W099 – Safety and Health in Construction Research Roadmap – Report for Consultation *CIB Publication 376*: CIB General Secretariat.

Gherardi, S., & Nicolini, D. (2002). Learning the trade: A culture of safety in practice. *Organization, 9*(2), 191–223.

Giang, D. T. H., & Pheng, L. S. (2011). Role of construction in economic development: Review of key concepts in the past 40 years. *Habitat International, 35*(1), 118–125.

Gidado, K. I. (1996). Project complexity: The focal point of construction production planning. *Construction Management and Economics, 14*(3), 213–225.

Global Construction Perspectives & Oxford Economics. (2013). *Global Construction 2025*. London, UK: Global Construction Perspectives and Oxford Economics.

Hampson, K., & Brandon, P. (2004). *Construction 2020: A Vision for Australia's Property and Construction Industry*. Brisbane, Australia: Cooperative Research Centre for Construction Innovation.

Health and Safety Executive. (2005). *A Review of Safety Culture and Safety Climate Literature for the Development of the Safety Culture Inspection Toolkit*. Bristol: HSE Books.

Health and Safety Executive. (2009). Phase 1 Report: Underlying causes of construction fatal accidents – A comprehensive review of recent work to consolidate and summarise existing knowledge (C. Division, Trans.). Norwich: Health and Safety Executive.

Health and Safety Executive. (2013). Health and Safety in Construction in Great Britain: Work-Related Injuries and Ill Health Retrieved 23 July 2014, from http://www.hse.gov.uk/statistics/industry/construction/construction.pdf

HM Government. (2013). *Construction 2025*. London: HM Government.

Holt, A. S. J. (2005). *Principles of Construction Safety*. Oxford: Blackwell Science.

Hopkins, A. (2005). *Safety, Culture and Risk: The Organisational Causes of Disasters*. Sydney: CCH.

International Labour Organization. (2003). *Safety in Numbers: Pointers for Global Safety Culture at Work*. Geneva: International Labour Organization.

Jackson, B. J. (2010). *Construction Management Jumpstart* (2nd ed.). Indianapolis: Wiley.

Kanji, G. K., & Wong, A. (1998). Quality culture in the construction industry. *Total Quality Management, 9*(4–5), 133–140.

Lingard, H., & Rowlinson, S. (2005). *Occupational Health and Safety in Construction Project Management*. Oxon: Spon Press.

Mayhew, C., & Quinlan, M. (1997). Subcontracting and occupational health and safety in the residential building industry. *Industrial Relations Journal, 28*(3), 192–205.

McDonald-Wilmsen, B. (2009). Development-induced displacement and resettlement: Negotiating fieldwork complexities at the Three Gorges Dam, China. *The Asia Pacific Journal of Anthropology, 10*(4), 283–300.

Ministry of Manpower. (2013). *Occupational Safety and Health Division Annual Report 2013*. Singapore: Ministry of Manpower.

Murie, F. (2007). Building safety-an international perspective. *International Journal of Occupational and Environmental Health, 13*(1), 5–11.

Ness, K. (2012). Constructing masculity in the building trades: 'Most jobs in the construction industry can be done by women'. *Gender, Work and Organization, 19*(6), 654–676.

NSW Government. (2010). Sydney Harbour Bridge Retrieved 7 November 2013, from http://sydney-harbour-bridge.bos.nsw.edu.au/

Peters, G. A., & Peters, B. J. (2006). *Human Error: Causes and Control*. Boca Raton: CRC Press.

Petersen, D. (1988). *Safety Management: A Human Approach* (2nd ed.). New York: Aloray.

Project Management Institute. (2013). *A Guide to the Project Management Body of Knowledge: PMBOK Guide* (5th ed.). Pennsylvania, USA: Project Management Institute.

Reason, J. (1990). *Human Error*. New York: Cambridge University Press.

Reason, J. (2000). Safety paradoxes and safety culture. *Injury Control and Safety Promotion, 7*(1), 3–14.

Reese, C. D., & Eidson, J. V. (2006). *Handbook of OSHA Construction Safety and Health* (2nd ed.). Boca Raton, FL: CRC Press.

Rockefeller Center. (2013). Art & History at Rockefeller Center Retrieved 7 November 2013, from http://www.rockefellercenter.com/art-and-history/

Safe Work Australia. (2013). Construction Fact Sheet Retrieved 23 July 2014, from http://www.safeworkaustralia.gov.au/sites/swa/about/publications/pages/fs2010constructioninformationsheet

Schwatka, N. V., Butler, L. M., & Rosecrance, J. R. (2012). An aging workforce and injury in the construction industry. *Epidemiologic Reviews, 34*(1), 156–167.

Stone, R. (2008). Three Gorges Dam: Into the unknown. *Science, 321*(5889), 628–632.

Wagner, P. (2010). Safety – A Wicked Problem. Retrieved 24 September 2014, from http://www.peterwagner.com.au/wp-content/uploads/Safety-A-Wicked-Problem2.pdf

2 Economics of Safety

This chapter discusses the economic aspects of safety in construction and engineering, including costs, benefits, return on investment (ROI), investment optimisation and evaluation. Although efforts have been made to improve safety, many barriers remain and those influenced by economic considerations are particularly dominant. The dynamic characteristics of the industry and the uncertainty of demand compel contractors to rely on casual labour and subcontractors (Kheni, 2008). This subcontracting practice has the potential to intensify safety risks as economic pressures and intense competition penalise those who try to do their work safely due to their higher tender prices (Mayhew & Quinlan, 1997). Limited financial capability is also constantly considered as one of the main barriers in implementing safety measures, particularly for small contractors (Loosemore & Andonakis, 2007). To further exacerbate the situation, the construction industry is also subjected to cyclical economic downturns (Dainty et al., 2001) and generally has a low and unreliable rate of profitability (Egan, 1998). As a result, decisions in relation to safety provisions may not be based upon ethical considerations and basic rights to safe workplace, but upon economics. This eventually leads to long hours of work, a low concern for safety and shortcuts in construction safety practices (Loosemore & Andonakis, 2007). In short, safety investments, that is, costs paid as a result of an emphasis being placed on safety (Hinze, 1997), have been considered as being expensive, but necessary only to avoid costly government fines (Linhard, 2005). As a result, when it comes to decision-making regarding the priorities between different project objectives in terms of time, cost, quality, safety and environmental impacts, often, investment made for safety improvement is perceived as non-return or low return by top management and boardroom members. Therefore, in this chapter we attempt to provide

Strategic Safety Management in Construction and Engineering, First Edition.
Patrick X.W. Zou and Riza Yosia Sunindijo.
© 2015 John Wiley & Sons, Ltd. Published 2015 by John Wiley & Sons, Ltd.

concrete evidence in terms of cost versus benefits of investment into safety management, so as to demonstrate to top management and decision-makers the worth of investing in safety. We also attempt to provide a safety investment optimisation model and a safety investment evaluation framework, to help top management make strategic decisions relevant to safety investment.

Costs of construction accidents

An accident can be defined as an unplanned event that results in injury or ill health of people, or damage or loss to property, plant, materials or the environment, or a loss of a business opportunity (Hughes & Ferrett, 2011). From a nationwide perspective, industrial accident costs have a significant impact on the national economy. In Australia, the cost of workplace injury and illness for the 2008–09 financial year was $60.6 billion, approximately 4.8% of the Australian GDP (Safe Work Australia, 2012a). The construction industry accounts for $6.4 billion or 10.6% of the total cost of workplace injury and illness. It is important to recognise that this cost is borne by workers, their families, the broader community, and employers (Safe Work Australia, 2012b). Similarly, from an organisational perspective, accidents can also be costly. Therefore, quantifying accident costs in monetary value becomes strategically important for business enterprises.

Cost classifications

Heinrich (1931) is perhaps the first who systematically studied accident costs. He classified accident costs into direct and indirect costs and these classifications are still widely used today. There are many interpretations on items that should be considered as direct and indirect costs. In general, direct costs include worker's compensation costs, damage to buildings or equipment, production loss of the injured worker, fines and legal expenses, sick pay and increase in insurance premiums. Indirect costs include items such as production delays, business loss due to delays, the recruitment and training of replacement staff, reputation loss, accident investigation time and any subsequent remedial action required, administration time and lower morale (Hughes & Ferrett, 2011).

Poon et al. (2008) provided a comprehensive classification of the costs of construction accidents as given here:

- *Financial costs of construction accidents*. Loss due to the injured persons; loss due to the inefficiency of the workers who have just recovered from injury and resumed work; loss due to medical expenses; loss of productivity of other employees; loss due to damaged equipment or plant; loss due to damaged materials or finished work; loss due to idle machinery or equipment; other costs.

- *Social costs of construction accidents.* These costs are calculated on the basis of the type and severity of accidents, the age of the injured person, and other related situations and conditions.
- *Human pain and suffering costs of construction accidents.* Employees compensation claims; common law claims such as loss of earnings; loss of earning capacity; loss of personal property, medical expenses and miscellaneous expenses; pain, suffering and loss of amenities; loss of society; loss of dependency; loss of accumulation of wealth; loss of earning of the immediate family members; loss of personal property; funeral expenses; loss of services; bereavement.

They also provided detailed cases examples and calculations of different types of costs due to different types of construction accidents.

It is important to identify the economic costs borne because it allows an understanding of the incentives for employers and regulators to provide a safe workplace. The classification structure for economic costs is based on the following six conceptual cost groups (Safe Work Australia, 2012b):

- *Production disturbance costs (PDC).* Costs incurred in the short term until production is returned to pre-incident levels
- *Human capital costs (HCC).* Long-run costs, such as loss of potential output, occurring after a restoration of pre-incident production levels
- *Medical costs (MEDC).* Costs incurred by workers and the community though medical treatment of workers injured in work-related incidents
- *Administrative costs (ADMINC).* Costs incurred in administering compensation schemes, investigating incidents and legal costs
- *Transfer costs (TRANC).* Deadweight losses associated with the administration of taxation and welfare payments, and
- *Other costs (OTC).* Include costs not classified in other areas, such as the cost of carers and aids and modifications.

These six cost groups could be further divided into many cost components. Some of the cost components are considered as indirect costs, which need to be estimated individually under different severity categories. Safe Work Australia (2012b) classifies incidents into five severity categories as follows:

- *Short absence.* A minor work-related injury or illness, involving less than five working days' absence from normal duties, where the worker was able to resume full duties.
- *Long absence.* A minor work-related injury or illness, involving five or more working days and less than 6 months off work, where the worker was able to resume full duties.
- *Partial incapacity.* A work-related injury or illness which results in the worker returning to work more than 6 months after first leaving work.

- *Full incapacity.* A work-related injury or disease, which results in the individual being permanently unable to return to work.
- *Fatality.* A work-related injury or disease, which results in death.

Cost calculation

Following the methods adopted by Safe Work Australia (2012b), the process of estimating total costs of accidents consists of seven steps as follows:

1. Identify the major categories of economic costs borne by economic agents (employers, workers and the community)
2. Determine the best source of measurement for each cost item
3. Define the levels of severity of injury or disease to differentiate between incidents with different cost structures
4. Identify which cost items apply to each severity category
5. Determine the number of incidents which fall into each severity category, and the average duration of time lost for a typical incident in each category
6. Calculate the average cost of a typical incident in each severity category by aggregating the typical costs associated with each cost item
7. Calculate the total cost of all work-related incidents by combining the typical cost of an incident with an estimate of the number of such incidents and aggregating over all classes of incidents.

Cost item details and their calculation methods

Some of the indirect cost items were estimated across all applicable severity categories due to the lack of available data relating to distribution by severity. For example, the overtime and over-employment costs were considered to be included in all severity categories, while legal costs are assumed to be included for full incapacity and fatal accidents only. The detailed cost components under each cost group and their distributions are summarised in Table 2.1. Cost item definitions are provided in Tables 2.2, 2.3, and 2.4 along with the methods and assumptions to generate estimates, which are based on the methodology used by Safe Work Australia (2012b). Although the methods and assumptions are based on the Australian context, the principles are applicable to other countries.

An example of cost calculation

Using the principles and methods explained above, we estimated the average cost associated with each direct and indirect cost item as shown in Table 2.5. Based on our calculations, the costs of construction injuries borne by employers, workers, and the community range from AU$3,344 for short absence to AU$1,686,819 for full incapacity injury. The cost of a full incapacity injury is higher than the cost of a fatal injury because more ongoing costs will be incurred

Table 2.1 Definition of different types of incidents and severity category

Conceptual group	Cost item	Borne by agent[a]	Direct or indirect cost[b]	Distribution by severity[c]
Production disturbance costs (PDC)	Cost of overtime and over-employment	E	I	All
	Employer excess payments	E	I	All
	Staff turnover costs	E	D	–
	Staff training and retraining costs	E	I	PI, FI, FT
	Loss of current income	W	I	–
	Compensation payments	C	I	LA, PI, FI, FT
Human capital costs (HCC)	Loss of future earnings	W	I	PI, FI, FT
	Loss of government revenue	C	I	LA, PI, FI, FT
	Social welfare payments for lost income earning capacity	C	D	–
Medical costs (MEDC)	Threshold medical payments	E	D	–
	Medical and rehabilitation costs	W	I	All
	Rehabilitation	C	D	–
	Health and medical costs	C	D	–
Administrative costs (ADMINC)	Legal fines and penalties	E	D	–
	Investigation costs	E	I	All
	Travel expenses	W	I	All
	Legal costs	W	I	FI, FT
	Funeral costs	W	D	–
	Inspection and investigation costs	C	D	–
	Travel concessions for full incapacitated workers	C	D	–
Transfer costs (TRANC)	Deadweight costs of welfare payments and tax losses	C	D	–
Other costs (OTC)	Carers costs	W	I	FI
	Aids and modifications	W	I	FI

[a] E=Costs borne by employer, W=Costs borne by worker and C=Costs borne by community
[b] D=Direct cost, I=Indirect cost
[c] SA=Short absence, LA=Long absence, PI=Partially incapacity, FI=Full incapacity, FT=Fatality, All=All severity categories

Table 2.2   Definitions, methods and assumptions of the cost items for employers

Cost category	Definition	Estimation
Cost of overtime and over-payment (PDC)	Proportion of overtime related to work-related injuries and wage of workers that would not be required if there were no work-related injuries.	Average weekly earnings $\times$ duration of absence in weeks $\times$ 0.4.[a]
Employer excess payments (PDC)	Portion of the costs of a claim required to be paid by the employer before workers' compensation provisions begin.	Average cost per day claim multiplied by 3.3 days.[b]
Staff turnover costs (PDC)	The costs to the employer associated with hiring new employees to replace injured or absent workers. This includes advertising costs and the costs associated with time spent in the recruitment process.	Turnover and recruitment costs are estimated to be equal in value to 26 weeks at average earnings less the amount 'brought forward' by incidents.
Staff training and retraining costs (PDC)	The costs to the employer associated with training existing staff and retraining new staff. This could arise both from legislative requirements as work-related incidents or simply the need to train staff with new skills as a result of increased responsibility or changed duties.	Average weekly earnings $\times$ 2.5.[c]
Medical threshold payments (MEDC)	Portion of workers' medical expenses to be met by the employers as part of employer excess provisions.	Average threshold medical payments, $500 in payments.
Legal fines and penalties (ADMINC)	Costs associated with successful prosecutions associated with proceedings initiated by workers' compensation authorities as a result of serious work-related incidents.	Average fine per conviction $\times$ number of convictions/total number of incidents.[d]
Investigation costs (ADMINC)	Costs associated with conducting an investigation into an accident and the administrative cost of collecting and reporting information on work-related accidents	Worker compensation expenditure relating to conducting investigations.

[a] For claims of longer duration or severity (such as permanent incapacity and fatality), the injured worker is assumed to be replaced after 8 weeks. The distribution of labour on-costs is based on data from ABS Major Labour Costs survey, and includes costs such as payroll tax and superannuation.

[b] Employer excess provisions differ between jurisdictions, both in terms of nature and period. The most common form of employer excess is 4 days, where the employer is liable for the costs associated with the first four days of a claim. However, some jurisdictions require no employer excess provisions. The weighted average of the excess period over each jurisdiction is 3.3 days. For severity category 1, the actual days lost are used in this calculation. For other categories, 3.3 days is used to proxy employer excess payments.

[c] Training and retraining are assumed to occupy approximately 2.5 weeks, covering both the time of the worker and also any training responsibilities of existing staff.

[d] Based on CPM estimates, the average fine per conviction is $27,595 and the prosecution rate is assumed to be 3% of incidents for permanent incapacity and 50% of incidents for fatalities.

Table 2.3   Definitions, methods and assumptions of the cost items for workers

Cost category	Definition	Estimation
Loss of current income (PDC)	Difference between pre-injury earnings and earnings following a work-related incident in the time following the incident to return to duties.	Residual item, total PDC less employer and society share of PDC.
Loss of future earnings (HCC)	Where the work-related injury or disease prevents natural career advancement and results in the worker being employed in a lower paid job, permanently incapacitated, or suffering a premature death.	Difference between expected future earnings in the absence of a work-related injury or disease and expected future income following the incident.[a]
Medical and rehabilitation costs (MEDC)	Expenditure on medical treatment not compensated via worker's compensation payments or government assistance.	The difference between medical costs incurred less medical payments covered by workers' compensation less government rebates.[b]
Travel expenses (ADMINC)	Expenditure for travel to doctors, rehabilitation centres, solicitors, and so on, less costs made in form of direct payments already included in the direct costs estimate.	Estimated from workers' compensation payments made for travel expenses (6% of non-compensation payments).
Legal costs (ADMINC)	Legal costs and expenses, less costs made in the form of direct payments already included in the direct costs estimate.	Difference between the average legal costs and overheads for a dispute and the amount received in compensation for legal cost.[c]
Funeral costs (ADMINC)	Real costs of bringing forward a funeral.	Average funeral costs are estimated at $3,617. Brought forward funeral costs are the discounted present value of a funeral at the time of life expectancy compared with the age at the time of the incident.
Carers (OTC)	For permanent cases only, the present value of future costs of carers.	Estimated applicable Disability Support Pension payments of $1,687 per annum, discounted to present value over the period between the incident and reduced life expectancy.

(continued overleaf)

Table 2.3 (*continued*)

Cost category	Definition	Estimation
Aids and modifications (OTC)	For permanent cases only, the present value of future costs for aids and modifications.	Estimated applicable Disability Support Pension payments of $530.4 per annum, discounted to present value over the period between the incident and reduced life expectancy.

[a] Workers are assumed to increase productivity (through experience and job knowledge) at the rate of 1.75% per annum. This figure is used in conjunction with discount and inflation rates to determine the present value of future income streams

[b] Medicare covered services that are bulk-billed are assumed to incur no cost to the individual. Workers are assumed to bear 15% of the total cost of the services when that service is not bulk-billed and covered by Medicare. On average, 47% of total costs result from Medicare covered services, with the remaining 53% of costs to be covered by private health insurance. Private health insurance covers 44% of cases, with the worker paying the gap payments of 5% on these costs. The costs of the remaining services are fully borne by the individual.

[c] Average legal costs and overheads per dispute are estimated to be $11,970 per dispute. It is estimated that disputes occur at a rate of 1 dispute per 8 claims. Average compensation for legal costs varies according to the severity of the incident, but comprises 62% of non-compensation payments.

by the employers, workers and the community after the occurrence of a full incapacity incident. The distribution of costs borne by each economic agent is as shown in Table 2.6.

Indirect costs

It should be noted that not all indirect costs are included in Table 2.5. For example, tarnished reputation and reduced competitiveness due to an accident are difficult, if not impossible, to be quantified in monetary terms. A study based on the Hong Kong construction industry attempted to calculate the social costs of accidents (Tang et al., 2004), where social costs were defined as the costs incurred by the society because additional resources are required to be utilised when construction accidents occur. Using 1414 accident data in 119 construction projects, the study found that the social costs in 1999, 2000 and 2001 were about US$101 million, US$69.5 million, and US$49.7 million respectively. They further estimated that for every extra $1 of social safety investments (safety investments made by contractors and government departments) made during 1999 to 2001, a reduction of $2.27 in social costs of construction accidents can be achieved (Tang et al., 2004). If we consider the losses attributed to death, pain, and suffering experienced by workers as well as the emotional and psychological impacts caused to family members and friends of the affected workers, then these indirect costs will increase significantly (Ikpe, 2009).

Table 2.4   Definitions, methods and assumptions of the cost items for community

Cost category	Definition	Estimation
Lost revenue (PDC/HCC)	The potential revenue lost when a worker suffers reduced earning capacity due to severe work-related incidents.	The taxation value of the present value of all future earnings over the period in which the individual is unable to work or that is lost through premature fatality.[a]
Social welfare payments (PDC/HCC)	Sickness and social welfare payments borne by the government for people with disabilities or the unemployed.	Average cost per recipient of social welfare programmes.[b]
Health and medical costs (MEDC)	Costs borne by the government through the provision of subsidised hospital, medical and pharmaceutical services.	Total Medicare costs that are not borne by the worker.
Rehabilitation (MEDC)	Expenditure on vocational education and training, special treatment, and so on.	Average cost of rehabilitation service (per recipient) reported by the Commonwealth Rehabilitation Service.[c]
Inspection and investigation costs (ADMINC)	Costs incurred by the agency responsible for conducting inspections and investigations.	Average cost per inspection reported by workers' compensation jurisdictions.
Travel concessions for permanently incapacitated (ADMINC)	Travel concessions and other allowances offered to permanently incapacitated workers.	Expenditure on travel costs by workers' compensation jurisdictions as a proxy for travel concessions.[d]
Transfer costs (TRANC)	Deadweight costs of welfare payments and tax losses	

[a] Based on average weekly earnings over the period of lost earnings, with an average taxation rate of 40%. Savings, inflation and productivity rates are also applied in determining the present value of future income streams. This total is split into short- and long-term costs. Short-term costs are incurred in the period between the incident and return to work, while long-term costs are incurred in the period following nominal return to work or replacement and retirement or to reduced life expectancy.

[b] Workers who suffer severe incidents are assumed to rely on the Disability Support pension (average cost per case is $10,659 p.a.) following a period of compensation (for compensated incidents).

[c] Workers who suffer a permanent incapacity are assumed to relay on the Commonwealth Rehabilitation Service (average cost per case is $3,362 p.a.) following the period of compensation (zero for non-compensated incidents).

[d] The community is assumed to match compensation payments for travel costs 1-1 with the individual, in effect assuming a 50 per cent travel concession for severely incapacitated workers.

Table 2.5   Estimated costs of injuries for the period of 2005–06 in Australia

Group	Item	Short absence	Long absence	Partial incapacity	Full incapacity	Fatality
Production disturbance costs (PDC)	Overtime and over-employment	132	2,120	18,133	35,447	3,610
	Employer excess payments	1,495	1,495	1,495	1,495	1,495
	Staff turnover	–	–	29,498	29,498	29,498
	Staff training and retraining	–	–	2,588	2,588	2,588
	Loss of current income	331	5,299	45,333	–	–
	Social welfare payment	–	10,659	29,636	29,636	29,636
Total PDC		1,958	19,572	126,682	98,662	66,826
Human capital costs (HCC)	Loss of future earnings	–	–	15,500	1,032,482	1,032,482
	Loss of government revenue	–	1,320	11,385	258,120	258,120
	Compensation and welfare payments for lost income- earning capacity	–	10,529	86,886	168,290	168,290
Total HCC		0	11,579	113,771	1,458,892	1,458,892
Medical costs (MEDC)	Threshold medical payments	500	500	500	500	500
	Medical and rehabilitation costs	462	2,187	12,091	12,097	6,111
	Rehabilitation	–	–	–	9,667	–
	Health and medical costs	305	656	2,055	1,452	367
Total MEDC		1,267	3,343	14,646	23,716	6,978
Admin. costs (ADMINC)	Legal fines and penalties	–	–	–	828	13,798
	Investigation costs	28	527	832	2,374	2,840
	Travel expenses	6	27	257	11,232	404
	Legal costs	–	–	–	11,970	11,970
	Travel concessions for incapacitated workers	–	–	–	5,616	–
Total ADMINC		34	554	1,089	32,020	32,629
Transfer costs		85	590	4,484	31,015	6,205
Other costs (OTC)	Deadweight costs of welfare payments and tax losses					
	Carer costs	–	–	–	32,345	–
	Aids and modifications	–	–	–	10,169	–
Total OTC		–	–	–	42,514	–
Total Costs		3,344	35,638	260,672	1,686,819	1,571,529

Note: All costs are in Australian dollar

Table 2.6   Distribution of costs borne by employer, worker and community

Severity category	Employer (%)	Worker (%)	Community (%)
Short absence	74	10	16
Long absence	15	15	70
Partial incapacity	21	24	55
Full incapacity	4	63	33
Fatality	3	67	30

A study in the USA found that the estimated costs of injuries in the construction industry in 2002 were US$11.5 billion, which account for 15% of the costs for all industries (Waehrer et al., 2007). The average cost per case of injury was US$27,000 while the average per-fatality cost was US$4 million. This amount included direct costs, indirect costs, and the quality of life costs. There are other studies that attempted to objectively calculate the costs of construction accidents. Differing results ensued because of the lack of standardisation in the methods adopted. However, the underlying message is always the same. Accidents are costly and detrimental to business. Some people tend to marginalise accident costs because they believe that those costs are insured. However, we need to realise that uninsured or indirect costs may form a significant proportion of the total costs of accidents and these costs generally are borne by employers. Figure 2.1 summarises the components of direct and indirect costs of accidents.

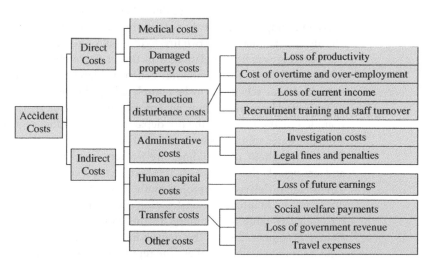

Figure 2.1   Accident costs

Indirect costs versus direct costs

Some studies have attempted to determine the ratio between indirect and direct costs of accidents. Heinrich (1931) estimated that indirect costs are four times greater than direct costs. Data collected from the US construction industry found that the ratio of indirect to direct costs for medical-case injuries is 4.2 and for restricted activity or lost-workday injuries it is 20.3 (Hinze & Appelgate, 1991). A study in the financial services sector in the UK found a ratio of about 3.3 (Monnery, 1998). Hughes and Ferrett (2011) suggested that the ratio ranges from 8 to as high as 36. These findings indicate that there is no generally accepted ratio between indirect and direct costs of accidents. The wide variance of this ratio may be attributed to different definitions of what should be considered as indirect and direct costs of accidents, the accuracy of survey methods, and specific conditions in different projects and regions (Feng, 2011). Company size may also contribute to the ratio difference. As compared to small companies, when an accident happens in large companies, more activities are initiated, more people are involved, more internal administrative processes have to be complied with, and more organisational levels have to be informed (Ikpe, 2009; Rikhardsson & Impgaard, 2004). This will inevitably increase the indirect costs of the accident and raise the ratio higher. Similarly, the ratio between indirect and direct costs of accidents tends to increase with the project size (Hinze & Appelgate, 1991). The main point here is that indirect costs of an accident could be significantly higher than its direct costs. Focusing too much on direct costs, which at first may seem to have relatively low impacts on the financial health of businesses, may fail to reveal the true losses to employers due to an accident (Feng, 2011).

As an example, consider the case of the Deutsche Bank fire in New York on 18 August 2007 that killed two fire-fighters and injured 115 other fire-fighters. The building was heavily damaged in the 11 September 2001 attack and had to be demolished. The demolition began in March 2007 and, before the fire happened, was scheduled to be completed by the end of 2008. The fire was likely started by workers smoking, ignoring the no-smoking rule on site. Compounded by the failure of the fire department to conduct regular inspections, a broken water supply system, a maze of sealed-off stairwells, and combustible debris throughout the building, fire-fighters faced major hurdles to control and extinguish the fire, which resulted in the two fatalities (Kugler, 2007). The accident delayed the project until it was finally completed in February 2011 (Lower Manhattan Construction Command Center, 2011). This delay has added roughly $100 million to the cost of rebuilding the World Trade Center (Topousis, 2010). The main contractor has agreed to pay about $15 million to the victims' families. The reputation of the contractor was seriously hit by the highly publicised accident, resulting in the loss of trust from the public and government (Smith, 2012). The subcontractor, hired to do the demolition job, was found guilty of reckless endangerment, a misdemeanour. The subcontractor filed for bankruptcy in 2012 (Palank, 2012).

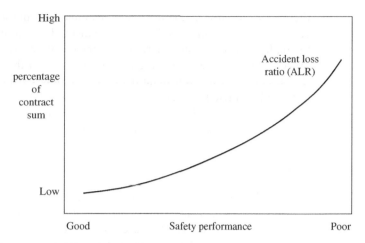

Figure 2.2   Accident loss ratio

Cost of accidents versus safety performance

The total accident cost on a construction project greatly depends on its safety performance. If the safety performance is good, the costs of accidents will be low, and vice versa. Figure 2.2 depicts the relationship between the two with the help of accident loss ratio (ALR), which is the ratio between the cost of accidents and the contract sum of the project (Poon et al., 2008).

$$\text{Accident loss ratio (ALR)} = \frac{\text{Total costs of accidents}}{\text{Contract sum}} \times 100\%$$

Benefits of investment in safety

First and foremost, safety is vital because of its impact on the wellbeing, health and lives of people. High rates of accidents and fatalities have caused much pain and suffering, conditions that cannot be justified by any means.

Tangible benefits

There are obvious tangible benefits of focusing on safety. The reduced rate of accidents will automatically reduce the costs incurred due to accidents. As discussed in the previous sections, an accident can have a very high cost and may force an organisation to shut down. Although insurance companies will cover part or all of the compensation costs, losses, and expenses incurred, the accident record will lead to a higher insurance premium (Li & Poon, 2013).

Fewer accidents also mean fewer interruptions to the production process, allowing project personnel to concentrate on important aspects of the project, such as meeting project objectives. When an employee is injured, not only is there a cost to provide treatment for the employee, but there is another added cost for recruiting and training a replacement. There is also lost revenue due to employee absenteeism where the degree of interruption to the production process is influenced by the degree of importance of the injured employee to the process. If, at the time of an injury, the employee happens to be employed in a highly contributory role, his or her absence will likely cause a more severe disruption to the production process than an injury to an employee who has a minor role (Rikhardsson & Impgaard, 2004).

Intangible benefits

Besides the tangible benefits, there are also intangible benefits of safety management implementation. Today, safety has become a social and moral responsibility. It is the right of every employee to go home safely every day and employees should not be treated as objects to achieve corporate goals. The reputation of an organisation is at stake when it does not implement proper measures to protect the safety and wellbeing of its employees, which will affect its tendering opportunities (Li & Poon, 2013; Lingard & Rowlinson, 2005). Accidents also reduce the morale of workers, while, on the contrary, accident prevention programmes strengthen morale, improve productivity and promote job satisfaction. Furthermore, a good safety record and proven safety management system are valuable marketing tools to attract new clients and support business expansion (Holt, 2005; Ikpe, 2009). All these tangible and intangible benefits will eventually contribute positively to the financial health and survivability of the organisation (Fernández-Muñiz et al., 2007; Ikpe, 2009).

Table 2.7 presents some tangible and intangible benefits that could be derived from investment in construction safety. The benefits are interrelated so that one tangible benefit may lead to other tangible and intangible benefits. For example, the use of personal protective equipment (PPE) will reduce the number of accidents. This will lead to other benefits, including a satisfied client because the project is completed on time. Client satisfaction will lead to increased reputation, increased market share and higher profit. Tables 2.8 and 2.9 provide indicators and methods, compiled from previous research, for measuring the tangible and intangible benefits respectively (Chi et al., 2005; Fernández-Muñiz et al., 2007; Hoonakker et al., 2005; Kartam et al., 2000; Ng et al., 2005; Rechentin, 2004; Rikhardsson & Impgaard, 2004; Teo et al., 2005; Vredenburgh, 2002).

Unmeasurable benefits of safety investment

In addition, there are intangible benefits that cannot be quantified. This is because these intangible benefits are affected by many variables and there are no suitable indicators that can be identified or significant enough to build a framework to measure these benefits. Examples of non-measureable benefits

Table 2.7 Tangible and intangible benefits of investment in construction safety

Investment in construction safety	Tangible benefits						Intangible benefits			
	Reduced number of accidents / injuries	Reduced cost of accident	Reduced number of interruption in production process	Increased market share	Higher profit	Better workers' motivation	Increased productivity	Improved reputation	Improved competitiveness	More satisfied clients
Product and equipment										
Personal Protective Equipment (PPE)	X	X								
Tools and machineries for working safely	X	X	X			X	X	X		
Regular maintenance of equipment	X	X	X			X	X	X		
Fall protection measures	X	X				X				
Safety management system										
Safety in design	X	X	X	X	X	X	X	X	X	X
Policy, rules and regulations	X	X							X	
Safety monitoring and consultation	X	X				X		X		
Monetary and non-monetary incentives for safety						X		X		
Set safety goals at different management levels and provide necessary resources	X	X						X		
Safety campaigns	X	X				X				

(continued overleaf)

Table 2.7 (*continued*)

Investment in construction safety	Tangible benefits					Intangible benefits				
	Reduced number of accidents / injuries	Reduced cost of accident	Reduced number of interruption in production process	Increased market share	Higher profit	Better workers' motivation	Increased productivity	Improved reputation	Improved competitiveness	More satisfied clients
Training										
Safety training for all employees	X	X								
Targeted training for individual sites	X	X	X							
Safety training for different types of jobs	X	X	X				X			
Behavioural-based training	X	X				X				

Table 2.8   Indicators and measurement methods for tangible benefits of investment in construction safety

Item	Measurement indicators	Calculation methods
Reduced number of accidents and injuries	Number of accidents; accident rate; amount of money put in safety investment which resulted in a reduced number of accidents and therefore improving safety performance	Calculate the number and differences of accidents before and after investment in safety
Reduced cost of accidents	Cost of accidents incurred before and after the implementation of safety investment	Calculate the costs and the differences of accidents before and after investment in safety
Reduced number of interruption in production process	Amount of interruption due to accidents; occupational accident cost (replacing employees who got involved in accidents); identify the lost revenue due to work time lost and employee absenteeism	Calculate the reduced cost of accidents (in relation to employee replacement and absenteeism) where safety investment is made.
Increased market share	Assess the market share of the organisation	Calculate the increase in the market share after making safety investment
Higher profit	Reduced cost of accidents and increased market share result in higher profit	Calculate the increase in profit after implementing safety investment

of investment in construction safety are (Loosemore & Andonakis, 2007; Törner & Pousette, 2009) as follows:

- Increase of the organisation's innovativeness
- Ability to maintain quality of workforce
- Improvement in safety knowledge
- Improvement in transfer of information and change of attitude towards safety

Return on investment in safety management

One way to assess the economic benefits of safety objectively is by measuring its ROI. ROI is a quantitative measure of performance used to evaluate the efficiency of an investment or to compare the efficiency of a number of different investments. If an investment does not have a positive ROI, or if there are other investments with a higher ROI, then the investment should not be undertaken

Table 2.9   Indicators and measurement methods for intangible benefits of investment in construction safety

Item	Measurement indicators	Measurement methods
Improved com-petitiveness	Reputation in safety; productivity; client satisfaction	Measure the increase in financial performance due to the indicators
Increased productivity	Productivity rate; number of accidents; reduced accident rate; reduced cost of accidents; safety behaviour	Measure the increase in productivity after the reduction in number of accidents
Improved reputation	Reputation and position in the market due to safety; market share	Measure the increase in market share and reputation
Satisfied clients	The achievement of project objectives	Measure the increase in terms of client satisfaction
Better workers' motivation	Down time and absenteeism; worker satisfaction rating; turnover rate	Measure employee satisfaction and the rates of turnover and absenteeism

(Burke, 2003). An ROI analysis can be used in different types of investment,s including safety in construction projects and business. ROI is expressed as a percentage or a ratio in relation to the gain from an investment and the cost of that investment, and is calculated using the following formula:

$$\text{Return on investment} = \frac{(\text{Gain from investment} - \text{Cost of investment})}{\text{Cost of investment}} \times 100\%$$

In order to determine the ROI in safety management in construction business/ projects, three sets of data are required. First, the costs of construction accidents to be borne by the employer, worker and community, who were considered as the beneficiaries of investing in safety, should be determined. An example to calculate the costs of accidents has been provided in earlier sections in this chapter.

Second, the total investment in safety in the organisation should be measured. This total investment is then compared to the construction industry average to determine the extra investment that has been put into the implementation of the organisation's safety programme. Safety investment often refers to costs of accident prevention activities, which can be classified into six main categories as follows (Feng, 2011):

1. *Safety staffing costs.* On-site staffing costs, head office staffing costs.
2. *Safety training costs.* Formal safety training courses, in-house safety training.

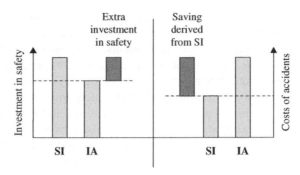

SI = Safety investment in the organisation
IA = Industry average

Figure 2.3   Calculation of investment in and savings from safety management in construction

3. *Safety equipment and facilities costs.* PPE, safety facilities (material, tools and machinery), safety facilities (manpower).
4. Costs of new technologies, methods or tools for safety.
5. *Safety committee costs.* Budget for safety committee, time lost due to safety committee activities.
6. *Safety promotion and incentive costs.* Safety promotion costs, safety incentive costs, costs relating to safe work method statements.

Third, the benefits, represented by the saving from the reduced number of accidents, as determined by the comparison between the organisation's safety performance and the construction industry average, should be measured. The difference between safety performance in the project/organisation and in the industry can be considered as the saving or gain from safety investments. Figure 2.3 illustrates this process. The case study presented in the next section gives a detailed example of how to calculate ROI in safety investment in construction projects/business.

The following section provides a case study to demonstrate how ROI in construction safety can be calculated step by step.

A case study on return on investment in safety risk management

Step 1 – calculate savings (gains) derived from safety investment

The Medical Research Centre (MRC) project in Australia is used as a case study to show how the ROI works in safety calculation. Site work in the MRC project was commenced in November 2007. The duration of the project was 30 months with 711,192 hours worked. The total project contract sum was approximately $100 million in which 3.02% was allocated to safety investments (the six categories of safety investments). This amount is higher than the industry average, which was assumed to be about 2%. Feng (2011) found that the average of basic safety investments in Singaporean construction industry is 1.59% of the project contract sum. Basic safety investments are investments required by industry or government regulations and construction processes to meet minimum safety standards. Organisations that focus on safety aim to exceed the minimum requirements by having additional voluntary safety investments. Feng (2011) found that the optimal level of voluntary safety investment is 0.44% of the contract sum. Therefore, we assume in our calculation that the average of safety investment in the construction industry is 2%, which is lower than the safety investment in the case study project (3.02%). Table 2.10 presents the statistics of injuries in the Australian construction industry for the period of 2007–09, while Table 2.11 presents the comparison of safety performance between the MRC project and the industry average.

In Table 2.11, the numbers for the industry average were estimated based on the statistics in Table 2.10. For example, the average frequency rate of long absence injury for the financial years 2007–09 is 8.25. Thereafter, the number of

Table 2.10   Statistics of injuries in the Australian construction industry in 2007–2009

Description		2007–2008	2008–2009	Average 2007–2009
Short absence[a]	Number of claims	5,454	5,520	5,487
	Frequency rate[b]	16.4	13.1	14.75
Long absence	Number of claims	11,560	11,709	11,634.5
	Frequency rate	8.6	7.9	8.25
Partial incapacity	Number of claims	1,730	1,838	1,784
	Frequency rate	1.3	1.2	1.25
Full incapacity	Number of claims	1,115	1,133	1,124
	Frequency rate	0.8	0.7	0.75
Fatality	Number of claims	37	42	39.5
	Frequency rate[c]	2.8	2.7	2.75

[a] Data for short absence injuries are based on NSW statistics because the project is located in NSW and jurisdictions in Australia have different methods to calculate short absence
[b] Frequency rate is the number of cases expressed as a rate per 1 million hours worked by employees. The formula for calculating frequency rate per million hours worked is:

$$\text{Frequency rate} = \frac{\text{Number of accidents in the period}}{\text{Total hours worked during the period}} \times 100$$

[c] Frequency rate for fatal injuries is based on 100 million hours worked
Source: Safe Work Australia (2012c)

Table 2.11 Comparison of safety performance between
the MRC project and industry average

Type	Number of injuries		
	MRC	Industry average	Difference
First aid injury	50	N/A	–
Short absence	2	9.28	7.28
Long absence	1	5.87	4.87
Partial incapacity	0	0.89	0.89
Full incapacity	0	0.53	0.53
Fatality	0	0.02	0.02

Table 2.12 Saving due to better safety performance in the MRC project

	Short absence	Long absence	Partial incapacity	Full incapacity	Fatality
Difference in safety performance	7.28	4.87	0.89	0.53	0.02
Costs of injuries ($)[a]	3,681	39,222	286,889	1,856,466	1,732,581
Saving ($)	26,794	191,011	255,331	983,927	34,592
Total saving ($)	1,491,654				

[a] See total costs in Table 2.5. The costs presented in Table 2.4 have been adjusted with 3.9% of inflation per year.

long absence injuries is measured using the following formula:

$$\text{No. of long absence} = \frac{\text{Frequency rate of long absence} \times \text{hours worked}}{1,000,000 \text{ hours}}$$

$$= \frac{8.25 \times 711,192 \text{ hours}}{1,000,000 \text{ hours}} = 5.87$$

Once the difference in the number of injuries and the cost of relevant injuries are determined, the amount of saving due to the reduced number of injuries can be calculated by multiplying the difference in safety performance and cost of injuries. The results of the calculations are presented in Table 2.12. In the MRC project, nearly $1.5 million is saved due to better safety performance than the industry average.

Step 2 – calculate the (additional) amount of investment into safety

Data from the main contractor revealed that the total amount of safety investment in the MRC project was $3,021,126, thus with a total budget of $100 million, the safety investment ratio (SIR) of the project is approximately

3.02% ($3.02 million/$100 million × 100%). Based on the SIR of 2% for the industry average, the extra safety investment in the MRC project is calculated as follows:

$$\text{Extra safety investment} = \$3,021,126 - (\$100,000,000 \times 2\%) = \$1,021,126$$

Safety investments comprise expenses for accident prevention activities. Feng (2013) identified seven components of safety investments which are summarised in Table 2.13.

Step 3 – calculate return on investment (ROI)

As such, the ROI on safety investment in the MRC project is 46.08%, according to the following calculation:

$$\text{Return on investment} = \frac{\text{Saving} - \text{Extra safety investment}}{\text{Extra safety investment}} \times 100\%$$

$$= \frac{1,491,654 - 1,021,126}{1,021,126} \times 100\% = 46.08\%$$

Step 4 – analysis and discussion

This case study demonstrates the economic benefit of safety. Although the SIR of the MRC project (3.2%) is significantly higher than the SIR of the industry average (2%), the project had better safety performance, resulting in nearly $1.5 million of saving due to the reduced number of injuries as compared to the industry average. It was estimated that the safety investment in the project generated a ROI of 46.08%. The high costs of injuries related to full incapacity and fatality would severely impact project financial performance when those injuries occur. An effective safety management system, which may seem to generate hefty expenses, could bring huge saving to a project by reducing the number of injuries. In the end, the initial safety investment is easily covered by the saving generated from better safety performance.

It should be noted that the intangible benefits of safety were not included in ROI calculation of the case study. As explained earlier, these intangible benefits may include employee motivation, client satisfaction, image and reputation, organisation's market share, and so on. The overall economic benefit of safety would be much more significant if the value of these intangibles can be calculated.

Furthermore, frequency rates used to calculate the ROI in safety are considered as lagging indicators of safety performance. They show performance of the past, that is, measuring performance on the basis of what has transpired. They are useful to reveal trends and to show the bottom lines of whether or not safety measures have been effective. They are, however, reactive and not useful as early warnings (Øien et al., 2011). Assessing safety performance on the basis

Table 2.13   Components of safety investments

Component	Description	Measurement
Staffing costs	• On-site staffing costs, such as safety managers, safety officers and safety supervisors • Head office staffing costs, such as safety director and safety coordinator	• Salaries paid to safety personnel • Salaries and the percentage of time spent on safety work on each project because personnel may have non safety-related work
Safety equipment and facility costs	Provided to protect workers from potential hazards on construction sites, such as personal protective equipment (PPE), safety barricades and other facilities that help the workers to carry out their work safely, as well as the manpower needed for the installation and maintenance of these facilities; equipment and facilities essential to carry out construction work are excluded	The costs of safety equipment and facilities include the purchase of equipment, materials, machines and tools, and the costs of manpower for the installation and maintenance of these facilities
Compulsory training costs	Safety training courses for project managers, safety training courses for foremen and supervisors, safety training courses for workers and safety training courses for operators and signal persons	The dollars paid for the external training institutes
In-house safety training costs	Safety tool box talk, emergency response and drills, first-aid procedures, safety workshops for supervisors and above, safety seminars and exhibitions, demonstrations of safe work procedures and others	The cost of lost productivity due to participation in these activities, which can be calculated based on the total number of participants, average hourly wages of the participants and duration and frequency of each in-house training activity
Safety inspection and meeting costs	Safety inspections and safety meetings which may consume the productive time of the participants and may cause the interruption of some ongoing construction work	The lost productivity due to the participation in the inspections and meetings and the interruption of ongoing construction work

(continued overleaf)

Table 2.13 *(continued)*

Component	Description	Measurement
Safety incentive and promotion costs	Safety incentive, safety promotion and safety awareness programmes	The expenses on the printing of pamphlets and posters, production of safety advertising boards and banners, organising of safety campaigns, financial support for safety committee activities, monetary rewarding of workers, management staff or subcontractors who achieve a good safety standard of work
Safety innovation costs	The use of new technologies, methods, procedures, or tools in order to improve safety performance of the project	Estimating the direct investments in obtaining the innovations (purchase of new tools or technologies, costs of research and development, and training costs) and possible increased production costs or lost productivities incurred by the use of these innovations

of the number or rate of reported accidents is also considered an unsound basis for comparison or investigation. The reason is that organisations that diligently report and investigate accidents are disadvantaged in comparison to careless organisations that do not always report accident occurrences. As a result, it is hard to motivate organisations to accurately report the number of accidents (Ng et al., 2005). To overcome these weaknesses, leading indicators should also be used together with lagging indicators to assess safety performance. Leading indicators provide performance feedback before an accident occurs. They require systematic and periodic checks to ensure that activities are carried out safely (Øien et al., 2011). Safety culture and safety climate are examples of leading indicators, and are discussed in Chapter 3.

Optimisation of investment in safety risk management

There are several methods provided by research for the calculation of costs, benefits and optimisation of investment in relation to safety risk prevention, mitigation and management, such as Laufer (1987), Brody et al. (1990), Tang et al. (2004), De Saram and Tang (2005), Hallowell (2010) and Feng (2011). In such calculations, the accumulation of costs to implement accident prevention activities is the amount of total safety investment, and the SIR can be calculated on the basis of the following formula:

$$\text{Safety investment ratio (SIR)} = \frac{\text{Total safety investment}}{\text{Contract sum of the project}} \times 100\%$$

As in the case of accident costs, a relationship can also be established between SIR and safety performance as depicted in Figure 2.4 (Poon et al., 2008). Figure 2.4 implies that when risk exposure is low, further safety investment may not be economical, that is, it may not be a worthwhile investment economically. This shows that in order to maximise the economic benefits of safety investment, there is an optimum value of safety that organisations should determine. This is the focus of discussion in this section.

Although logically safety investment should positively influence safety performance, this influence is largely an issue of probabilities. If safety investment is high, the probability of an accident occurring becomes relatively small, while if the safety investment is low, the probability of accident occurring becomes relatively high. However, there is always a possibility that there might be no accident even if there is no safety investment, while a fatality might happen even though there is a considerable amount of safety investment (Feng, 2011).

Following from the above probability statement, it appears that the economic law of diminishing returns is also applicable in safety investments. The law states that when at least one factor of production is fixed, successive increases in the input of a variable factor eventually yield smaller and smaller increments in output (Frank et al., 2009). Malthus (1798), in his famous treatise, *An Essay on the Principle of Population*, explained that land, as a factor of production, is in limited supply and would lead to diminishing returns. In order to increase output from agriculture, farmers would have to farm less fertile land or farm with more intensive production methods. In both cases, however, the returns

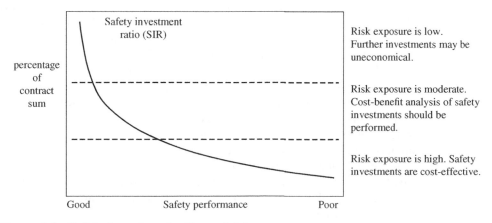

Figure 2.4   Safety investment ratio and risk exposure

from agriculture would diminish over time. This law, first thought to apply only to agriculture, has been accepted as an economic law underlying all production enterprises. This law basically suggests that as a system approaches perfection, market saturation or the natural environment will constrain the effectiveness of the production process, causing the output derived from an input to fall with increasing investments of other inputs (Feng, 2011).

According to the law of diminishing returns, given a certain level of inputs in activities to improve safety performance and a certain level of inherent hazard level of the project, each additional unit of input will yield less and less output, that is, the improvement of safety performance. Figure 2.4 can be used to visualise the applicability of the law of diminishing returns in safety investments. When the level of input is relatively small, small increments of safety investments will add substantially to output, which is the reduction of risk exposure. As the level of safety investments increases, the law of diminishing returns applies. Too many investments to improve safety performance will make part of the investments less effective and cause the product of safety investments to fall. Eventually some investments may be deemed to be uneconomical (Lingard & Rowlinson, 2005).

As mentioned earlier, Feng (2013) conducted a study on optimum safety investments in the Singaporean construction industry. He classified safety investments into basic safety investments (about 1.59% of the contract sum), which are expenses on accident prevention activities that are required by industry or government regulations to meet minimum safety standards, and voluntary safety investments (about 0.44% of the contract sum), which are expenses on accident prevention activities that are generally determined by individual companies or projects. He found that voluntary safety investments are more effective in improving safety performance than basic safety investments. The stronger relationship between voluntary safety investments and safety performance shows that safety investments would be more effective when construction organisations choose to invest in areas deemed crucial, based on the specific needs of individual projects. This essentially means that safety investments need to be contextualised, a principle that has also been supported by the results of many studies (Aksorn & Hadikusumo, 2008; Findley et al., 2004; Jaselskis et al., 1996; Poon et al., 2008).

Further analysis shows that the optimal level of voluntary safety investments varies depending on the levels of safety culture and project hazard condition. The highest level of optimal voluntary safety investment occurs with the highest project hazard level and lowest project safety culture level, while the lowest level of optimal voluntary safety investment occurs with the lowest project hazard level and highest project safety culture level. It means that the optimal level of voluntary safety investment tends to decline with the increase of safety culture level when holding the project hazard level constant. Therefore, fostering a strong safety culture will not only improve safety performance but also contribute to lowering the expenditures on safety implementation (Feng, 2011). Safety culture is discussed further in Chapter 3.

Although Feng (2011) found that the optimal level of voluntary safety investments is about 0.44% of the contract sum, it does not mean that this figure should be considered as the maximum amount of voluntary safety investments. Figure 2.5 describes the relationships among voluntary SIR, ALR, total safety costs ratio (TCR) and accident frequency rate (AFR). As the voluntary SIR is increased, the AFR declines. On the other hand, the ALR curve has a positive slope because the total accident costs vary positively with the AFR. The TCR curve is the vertical sum of the voluntary SIR curve (related to Figure 2.4) and the ALR curve (related to Figure 2.2; thus there should be a minimum point on the TCR curve, that is, the point Min in Figure 2.5. Coinciding with the minimum level of total safety costs, from the financial perspective, $y1$ represents the potential accident cost and $y2$ represents the optimum level of voluntary safety investments;both are expressed by the percentage of the project contract sum. Based on the finding, a safety investment at the optimum level ($y2$) would result in the best financial performance and a fairly good safety performance. It means that safety investments which are less than $y2$ would

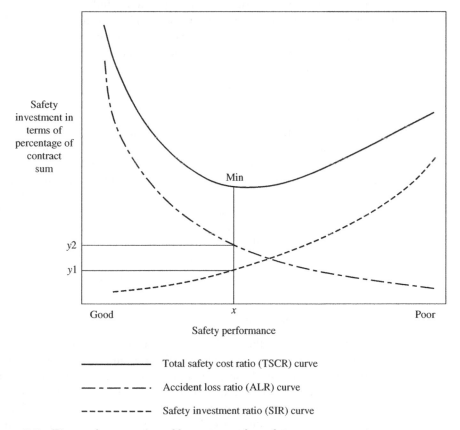

Figure 2.5   **The optimum point of investment in safety management**

result in both financial losses and poorer safety performance, while safety investments which are greater than $y2$ would also result in higher cost, but would potentially achieve a better safety performance. Therefore, $y2$ represents not only the financially optimum level of voluntary safety investments, but also the minimum level of voluntary safety investments. The potential benefits of better safety performance, particularly its intangible and hidden benefits, may outweigh the increase of costs resulting from voluntary safety investments greater than the optimum level (Feng, 2011).

Evaluation of investment in safety risk management

In this section we propose the use of econometrics as a method to evaluate the effectiveness of investment made for improving construction safety. A growing body of empirical research has been focused on the relationship between the costs and benefits of investing in safety in the construction industry and some of these research studies have been discussed in the previous sections in this chapter. However, these previous studies mainly describe the accident occurrence in terms of actual numbers, which were based on the hypothesis that when there is an investment, there will be an effect on construction safety performance and hence the number of accidents will be reduced. In reality, investment on construction safety does not guarantee reducing the actual number of accidents occurring in a construction project. The occurrence of accidents sometimes is out of control and unpredictable and, hence, remains random. This presents a gap between the existing theory and the actual practice, which means there is a need to develop a theoretical framework that can closely model accident randomness. This section presents an econometric analysis framework for evaluating the effectiveness of safety investment in construction projects based on the abovementioned 'reality'. In this framework the fundamental assumption is that the investment on safety can only reduce the probability of accidents from happening and the number of accident eventuation will behave as a random number in a range with some distributions, or, statistically speaking, shift the distribution of accidents. As the accident occurrence is considered to be a random number, the return on safety investment will automatically be a random number.

Econometric theory and its application

Econometrics is based on the development of statistical methods for estimating economic relationships, testing economic theories and evaluating and implementing government and business policy (Wooldridge, 2013). Econometric analysis is concerned with the quantitative relationships between economic variables and it can provide an important input to a manager's decision making. Typically, econometrics differs from other aspects of management science in that it considers problems primarily, though not exclusively, from a background of economics rather than of other disciplines (Ball & Burns, 1974). Econometrics is a good tool to analyse the relationship between variables and dependent variables.

There have been studies on safety using econometric theories and methods. Borooah et al. (1998) did an econometric analysis based on the injury compensation data in Queensland, Australia, and found that modelling workplace injuries using econometric analysis offers parameter and elasticity estimates applicable to policy making on workplace health and safety. Eluru and Bhat (2007) used an econometric analysis in the context of road safety. They examined the effects of factors such as driver characteristics, vehicle characteristics, roadway design attributes, environmental factors and crash characteristics to explain seat-belt use and crash-related injury severity levels. Their results show that unsafe drivers who do not wear seat belts are those likely to be involved in high injury severity crashes because of their unsafe driving habits. As such, when they are apprehended in the act of not wearing seat belts, they should be subjected to both a fine to increase the chances that they wear seat belts and mandatory enrolment in a defensive driving course to change their aggressive driving behaviour.

The economic benefit of construction safety is generally based on the assumption that there is a direct cause–effect relationship between investment in safety and safety performance improvement, that is, if there is an increase in safety investment, automatically there will also be an improvement of safety performance. However, in reality, this is not always the case as there is no guarantee of the direct relationship described above. Therefore, the fundamental assumption of econometrics in this case is that the number of accidents will act as a random number with some distributions, and safety investment can only reduce the probability of accidents or, statistically speaking, shift the distribution of accidents. Thus, this model should describe the distribution of probabilities of accidents on the basis of variable causal factors.

LOGIT regression model

In econometric research, a LOGIT model, as shown in Figure 2.6, is commonly used in modelling probability. When a conventional bivariate statistical model cannot elaborate the problem properly, the multinomial Logit model has been applied in many studies. A multivariate model is introduced to analyse the effects of safety investment and other variables on accident-occurring probability. By plotting the histogram of accidents and estimating its density function, the simulation of accidents can be conducted.

The outcome of the LOGIT model is always positioned between 0 and 1; thus it is appropriate for modelling probability. The independent variables are included in the right-hand side of the equation (see below). Ideally, all factors that influence the probability of accident occurrence should be identified and included, and safety investment is one of those factors. By investigating the coefficient of safety investment, its effectiveness in reducing the probability of

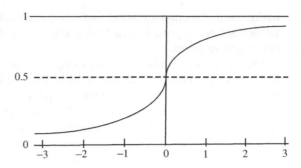

Figure 2.6   LOGIT model

accident occurrence can be determined (Zou et al., 2010). The formula of the LOGIT model is:

$$P(y_n = 1|x) = \Lambda(\beta_0 + \beta_1 x_1 + \beta_2 x_2 + \cdots + \beta_k x_k)$$

$$\text{where } \Lambda(z) = \frac{\exp(z)}{[1 + \exp(z)]}, \quad n = 1, 2, 3, \ldots$$

x_k are independent variables that influence the probability of accident occurrence

β_k are the independent variable coefficients, indicating the marginal effect of the independent variables.

From the above formula it can be seen that the LOGIT model allows transformation of its independent variable, keeps probability behaviour always between 0 and 1, rather than describe a linear relationship between dependent and independent variables, and so it is appropriate for modelling the probability. By investigating the coefficient of investment in the safety risk management programme and plotting the histogram of the number of accidents that occurred with and without investment in the safety risk management programme, and estimating its equation, the distribution of the accidents reduced can be estimated. Hence, the effectiveness of safety investment can be calculated using the formula.

Independent variables involved in construction accidents

As described above, the independent variables will be included in the right hand side of the LOGIT model/equation. Ideally, it should have all information which has significant effects on the probability of accident occurrence. Of course, the different types of investment in the safety risk management programme will be the significant factors, and the other elements from organization, society, workers, projects, and so on are also considered and analysed. Based on previous studies, we have selected a series of variables for the LOGIT model (listed in Table 2.14). Based on that list, the LOGIT econometrics model for analysing the effect of variables on accidents may be described as follows:

$$P(y_n = 1|x) = \Lambda[\alpha + (\beta_1 SSC + \beta_2 STC + \beta_3 SEC + \beta_4 SYC + \beta_1 SCC + \beta_1 SPC)]$$

$$+(\gamma_1 AW + \gamma_2 EW + \gamma_3 GW + \gamma_4 NW) + (\delta_1 SO + \delta_2 WO)$$

$$+(\varepsilon_1 SC + \varepsilon_2 WE + \varepsilon_1 SL) + (\theta_1 PZ + \theta_1 CS) + (\mu_1 GP + \mu_2 CP)$$

where $\Lambda(z) = \dfrac{\exp(z)}{[1 + \exp(z)]}$

Table 2.14   Factors that influence the probability of construction accident occurrence (Zou et al., 2010)

Categories	Independent variables	Measurement
Safety investment	Safety staffing costs (SSC)	On-site and head office staffing costs
	Safety training costs (STC)	Internal and external safety training costs
	Safety equipment and facility costs (SEC)	Costs for PPE, safety barricades, safety nets, and so on.
	Safety innovation and technology costs (SYC)	Online reporting tools and safety system development
	Safety committee costs (SCC)	Costs for safety meetings and inspections
	Safety promotion costs (SPC)	Safety posters, safety awards and safety awareness programmes
Worker profile	Average age (AW)	-
	Experience (EW)	Years of experience
	Gender (GW)	Gender balance
	Number of workers on site (NW)	-
Organisation	Size of company (SOC)	Number of employees or amount of revenue
	Worker supervision (WS)	Number of supervisors
Workplace	Site condition (SC)	Weather and geographical conditions
	Work environment (WE)	Lighting, noise, temperature, and so on.
	Site layout (SL)	Constrained or free access
Construction	Project size (PS)	Size and type of project, contract price
	Construction schedule (CS)	Tight or not, contract period
Policy	Government policy (GP)	Government expenditure on public safety
	Contract provisions (CP)	Compensation policy, insurance premiums

After LOGIT regression, the coefficient of each factor will show the effect of each variable on accident occurrence. The marginal effect of each category of safety spending according to different types of accidents can be represented by its coefficient. For example, the marginal effect of safety staff cost can be calculated by β_1. The marginal effect of other factors can be determined in the same manner, and hence the most efficient safety spending can be determined by its sensitivity. Furthermore, by plotting the histogram of the number of accidents that happened with and without some types of safety investment, the probability density function and cumulative probability function of accidents can also be estimated. And once the probability density function and cumulative probability function of accidents are determined, according to the ROI formula, the effectiveness of the safety investment can be calculated. Through the coefficient of each variable, the factor that has the most significant impact on accidents can be determined and the information can be useful for decision-makers in construction firms on safety investment.

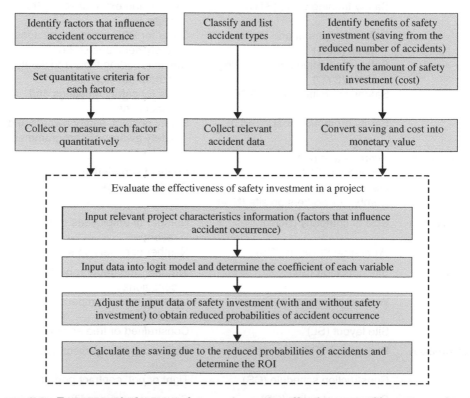

Figure 2.7   Econometric framework to evaluate the effectiveness of investment in safety risk management

A framework for evaluating investment in safety risk management

The econometric method can be integrated with the ROI model to establish a framework to evaluate the effectiveness of safety investment. Figure 2.7 explains the process.

It should be noted that real project safety data are needed in order to perform the calculation and evaluation framework and process discussed in the previous sections.

Conclusions

Safety investment is obviously not where revenues are generated, but it does generate profit, albeit indirectly, by reducing safety risks and, subsequently, the potential for loss. Commitment from contractors is needed to improve safety in the construction industry. However, commitment from clients, who have the economic power to facilitate safety implementation, is equally important, especially to support safety implementation in small- and medium-sized projects where competition for lowest price is more intense. Construction clients should realise that without their support, contractors will be heavily constrained in implementing safety measures, especially given the competitive nature of the industry. This chapter has provided evidence on the economic benefits of investing in construction safety by demonstrating that safety is not only the 'right thing' to do but also profitable and should be integrated into strategic business decisions. This chapter has further provided methods and techniques for optimising and evaluating the investment in construction safety risk management programmes.

It should be noted that although the costs, benefits, and ROI in safety may be measured in economic terms, the fundamentals of safety are about the preservation of human life and the protection of the human right for a safe working environment, and it may not be appropriate to measure human life in monetary terms. Therefore, making strategic safety management decisions should be based on not only economical modelling, but also the belief in the importance of safety and the ethics of fundamental human right. Essentially, these decisions should be made from moral, ethical and human right stand-points.

References

Aksorn, T., & Hadikusumo, B. H. W. (2008). Critical success factors influencing safety program performance in Thai construction projects. *Safety Science, 46*(4), 709–727.

Ball, R. J., & Burns, T. (1974). Econometric analysis and managerial decision making. *Omega, The International Journal of Management Science, 2*(3), 295–311.

Borooah, V. K., Mangan, J., & Hodges, J. (1998). Determinants of workplace injuries: An econometric analysis based on injuries compensation data for Queensland. *Economic Analysis and Policy, 28*(2), 149–168.

Brody, B., Létourneau, Y., & Poirier, A. (1990). An indirect cost theory of work accident prevention. *Journal of Occupational Accidents, 13*(4), 255–270.

Burke, R. (2003). *Project Management: Planning and Control Techniques* (4th ed.). Chichester, UK: John Wiley & Sons.

Chi, C.-F., Chang, T.-C., & Ting, H.-I. (2005). Accident patterns and prevention measures for fatal occupational falls in the construction industry. *Applied Ergonomics, 36*(4), 391–400.

Dainty, A. R. J., Millet, S. J., & Briscoe, G. H. (2001). New perspectives on construction supply chain integration. *Supply Chain Management: An International Journal, 6*(4), 163–173.

Egan, J. (1998). *Rethinking Construction*. London: Department of Trade and Industry.

Eluru, N., & Bhat, C. R. (2007). A joint econometric analysis of seat belt use and crash-related injury severity. *Accident Analysis and Prevention, 39*(5), 1037–1049.

Feng, Y. (2011). *Optimizing Safety Investments for Building Projects in Singapore*. PhD, National University of Singapore, Singapore.

Feng, Y. (2013). Effects of safety investments on safety performance of building projects. *Safety Science, 59*, 28–45.

Fernández-Muñiz, B., Montes-Peón, J. M., & Vázquez-Ordás, C. J. (2007). Safety culture: Analysis of the causal relationships between its key dimensions. *Journal of Safety Research, 38*(6), 627–641. doi: 10.1016/j.jsr.2007.09.001.

Findley, M., Smith, S., Kress, T., Petty, G., & Enoch, K. (2004). Safety program elements in construction: Which ones best prevent injuries and control related workers' compensation costs? *Professional Safety, 49*(2), 14–21.

Frank, R. H., Jennings, S., & Bernanke, B. S. (2009). *The Principles of Microeconomics* (2nd ed.). North Ryde, NSW, Australia: McGraw-Hill Australia.

Hallowell, M. (2010). Cost-effectiveness of construction safety programme elements. *Construction Management and Economics, 28*(1), 25–34.

Heinrich, H. W. (1931). *Industrial Accident Prevention: A Scientific Approach*. New York: McGraw-Hill.

Hinze, J. (1997). *Construction Safety*. New Jersey: Prentice-Hall.

Hinze, J., & Appelgate, L. L. (1991). Costs of construction injuries. *Journal of Construction Engineering and Management, 117*(3), 537–550.

Holt, A. S. J. (2005). *Principles of Construction Safety*. Oxford: Blackwell Science.

Hoonakker, P., Loushine, T., Carayon, P., Kallman, J., Kapp, A., & Smith, M. J. (2005). The effect of safety initiatives on safety performance: A longitudinal study. *Applied Ergonomics, 36*(4), 461–469.

Hughes, P., & Ferrett, E. (2011). *Introduction to Health and Safety in Construction* (4th ed.). Oxon, UK: Routledge.

Ikpe, E. O. (2009). *Development of Cost Benefit Analysis Model of Accident Prevention on Construction Projects*. PhD, University of Wolverhampton, Wolverhampton.

Jaselskis, E. J., Anderson, S. D., & Russell, J. S. (1996). Strategies for achieving excellence in construction safety performance. *Journal of Construction Engineering and Management, 122*(1), 61–70.

Kartam, N. A., Flood, I., & Koushki, P. (2000). Construction safety in Kuwait: Issues, procedures, problems, and recommendations. *Safety Science, 36*(3), 163–184.

Kheni, N. A. (2008). Impact of Health and Safety Management on Safety Performance of Small and Medium-sized Construction Business in Ghana. PhD, Loughborough University, Loughborough.

Kugler, S. (2007). Deutsche Bank Fire Probe: Many Failures Retrieved 9 January, 2014, from http://www.washingtonpost.com/wp-dyn/content/article/2007/08/28/AR2007082800272.html.

Laufer, A. (1987). Construction accident cost and management safety motivation. *Journal of Occupational Accidents, 8*(4), 295–315.

Li, R. Y. M., & Poon, S. W. (2013). *Construction Safety*. Heidelberg: Springer.

Lingard, H., & Rowlinson, S. (2005). *Occupational Health and Safety in Construction Project Management*. Oxon: Spon Press.

Linhard, J. B. (2005). Understanding the return on health, safety and environmental investments. *Journal of Safety Research, 36*(3), 257–260.

Loosemore, M., & Andonakis, N. (2007). Barriers to implementing OHS reforms – The experiences of small subcontractors in the Australian construction industry. *International Journal of Project Management, 25*(6), 579–588.

Lower Manhattan Construction Command Center. (2011). 130 Liberty Street Retrieved 9 January, 2014, from http://www.lowermanhattan.info/construction/project_updates/130_liberty_street__77170.aspx

Malthus, T. R. (1798). *An Essay on the Principle of Population*. London: J. Johnson.

Mayhew, C., & Quinlan, M. (1997). Subcontracting and occupational health and safety in the residential building industry. *Industrial Relations Journal, 28*(3), 192–205.

Monnery, N. (1998). The costs of accidents and work-related ill-health to a cheque clearing department of a financial services organisation. *Safety Science, 31*(1), 59–69.

Ng, S. T., Kam, P. C., & Skitmore, R. M. (2005). A framework for evaluating the safety performance of construction contractors. *Building and Environment, 40*(10), 1347–1355.

Øien, K., Utne, I. B., & Herrera, I. A. (2011). Building safety indicators: Part 1 - theoretical foundation. *Safety Science, 49*(2), 148–161.

Palank, J. (2012). Who is John Galt? Retrieved 9 January, 2014, from http://blogs.wsj.com/bankruptcy/2012/07/13/who-is-john-galt/

Poon, S. W., Tang, S. L., & Wong, F. K. W. (2008). *Management and Economics of Construction Safety in Hong Kong*. Hong Kong: Hong Kong University Press.

Rechentin, D. (2004). Project safety as a sustainable competitive advantage. *Journal of Safety Research, 35*(3), 297–308.

Rikhardsson, P. M., & Impgaard, M. (2004). Corporate cost of occupational accidents: An activity-based analysis. *Accident Analysis and Prevention, 36*(2), 173–182.

Safe Work Australia. (2012a). *The Australian Work Health and Safety Strategy 2012–2022*. Canberra: Safe Work Australia.

Safe Work Australia. (2012b). *The Cost of Work-related Injury and Illness for Australian Employers, Workers and the Community: 2008–09*. Canberra: Safe Work Australia.

Safe Work Australia. (2012c). Statistics Retrieved 10 January, 2014, from http://www.safeworkaustralia.gov.au/sites/swa/statistics/pages/statistics.

De Saram, D. D., & Tang, S. L. (2005). Pain and suffering costs of persons in construction accidents: Hong Kong experience. *Construction Management and Economics, 23*(6), 645–658.

Smith, G. B. (2012). Widow of Firefighter Joseph Graffagnino Reaches $10M Settlement 5 years After Hero Husband Dies in Deutsche Bank Fire Retrieved 9 January, 2014, from http://www.nydailynews.com/new-york/widow-firefighter-joseph-graffagnino-reaches-10m-settlement-5-years-hero-husband-dies-deutsche-bank-fire-article-1.1081291.

Tang, S. L., Ying, K. C., Chan, W. Y., & Chan, Y. L. (2004). Impact of social investments on social costs of construction accidents. *Construction Management and Economics*, *22*(9), 937–946.

Teo, E. A. L., Ling, F. Y. Y., & Chong, A. F. W. (2005). Framework for project managers to manage construction safety. *International Journal of Project Management*, *23*(4), 329–341.

Topousis, T. (2010). WTC Stalled Extra Year Retrieved 9 January, 2014, from http://nypost.com/2010/04/03/wtc-stalled-extra-year/

Törner, M., & Pousette, A. (2009). Safety in construction - a comprehensive description of the characteristics of high safety standards in construction work, from the combined perspective of supervisors and experienced workers. *Journal of Safety Research*, *40*(6), 399–409.

Vredenburgh, A. G. (2002). Organizational safety: Which management practices are most effective in reducing employee injury rates? *Journal of Safety Research*, *33*(2), 259–276.

Waehrer, G. M., Dong, X. S., Miller, T., Haile, E., & Men, Y. (2007). Costs of occupational injuries in construction in the United States. *Accident Analysis and Prevention*, *39*(6), 1258–1266.

Wooldridge, J. M. (2013). *Introductory Econometrics: A Modern Approach* (5th ed.). Mason, OH: South-Western.

Zou, P. X. W., Shi, V. Y., & Li, Z. (2010). An econometric evaluation framework for investment in construction safety. Paper presented at the 26th Annual ARCOM Conference, Leeds, UK.

3 Safety Climate and Culture

This chapter discusses key concepts and frameworks related to safety climate, safety culture, safety culture maturity and their measurement. It also presents several case studies to demonstrate the best international practice in implementing safety programmes for fostering strong safety culture in construction business and projects.

The term safety culture has become a popular catchphrase among high-risk industrial sectors, such as aviation, mining, agriculture and construction. Many organisations consider that developing safety culture is a must for preventing accidents and improving safety. However, despite its popularity, details about the definition, development and management of safety culture remain largely unclear. Different organisations have different interpretations of safety culture and different approaches to implement the notion in practice. Cox and Flin (1998) strongly cautioned that safety culture has become a generic solution for all psychological and human factor issues which may actually have exceeded the evidence for its utility. Confusion on what safety culture actually means has been exemplified in recent studies (Antonsen, 2009; Choudhry et al., 2007; Haukelid, 2008). To complicate the matter further, there is another concept called safety climate, which at the first glance seems to have the same nature and functions as that of safety culture. As a result, the terms safety culture and safety climate are often used interchangeably, blurring the border between the two.

This chapter first attempts to clarify the differences between safety climate and safety culture, followed by presenting an instrument to measure safety climate. Next, a framework is proposed to guide construction organisations to develop and manage their safety culture, which includes three aspects – psychological, behavioural and corporate. This chapter also includes safety culture maturity assessment criteria to help construction organisations understand

Strategic Safety Management in Construction and Engineering, First Edition.
Patrick X.W. Zou and Riza Yosia Sunindijo.
© 2015 John Wiley & Sons, Ltd. Published 2015 by John Wiley & Sons, Ltd.

their safety culture maturity level. Finally, the chapter presents three case studies demonstrating strategies that can be used in practice for fostering safety culture.

Safety climate

Definitions of safety climate

The origin of safety climate perhaps can be traced back to the idea of organisational climate coined in the 1930s. Thereafter, research on organisational climate evolved into studies on the perceptions of the workforce towards the social and managerial aspects of the work environment. Today the concept of organisational climate is widely accepted as an indicator of organisational effectiveness (Cox & Flin, 1998). The term safety climate itself was first used by Zohar (1980) when he conducted a study to measure the safety climate of Israeli industrial organisations. He defined safety climate as 'a summary of molar perceptions that employees share about their work environments (p. 96)'. Zohar argued that these perceptions serve as a frame of reference in determining employees' behaviour. As such, he concluded that managements need to change their attitudes and demonstrate their safety commitment in order to improve safety in the organisation. He also suggested that a safety climate questionnaire survey is a practical tool to compare safety performance between organisations because it is independent of factors such as technologies and risk levels that have caused difficulties in measuring safety performance in the past. Zohar's work has gained wide recognition and many research studies were conducted to further investigate the concept of safety climate. These studies have generated different safety climate definitions as follows:

- A set of perceptions or beliefs held by an individual and/or group about a particular entity (Brown & Holmes, 1986).
- Molar perceptions people have of their work settings (Dedobbeleer & Béland, 1991).
- The objective measurement of attitudes and perceptions towards occupational health and safety issues (Coyle et al., 1995).
- A summary concept describing the safety ethic in an organisation or workplace which is reflected in employees' beliefs about safety and is thought to predict the way employees behave with respect to safety in that workplace (Williamson et al., 1997).
- A construct that captures employees' perceptions of the role that safety plays within the organisation [and] a descriptive measure reflecting the workforce's perception of, and attitudes towards, safety within the organisational atmosphere at a given point in time (Mohamed, 2002).
- Shared employee perceptions of how safety management is being operationalised in the workplace, at a particular moment in time (Cooper & Phillips, 2004).

- Shared perceptions of employees about the safety of their work environment (Hahn & Murphy, 2008).
- An aspect of the organisation which is influenced by the way people behave, how they think and feel about safety issues (Loughborough University, 2009).

The list of the definitions above is not exhaustive, but there are similarities that we can draw from. First, all the definitions basically concur that safety climate is about employees' perceptions and attitudes towards safety in the organisation or in their workplace. Second, safety climate measures these perceptions and attitudes at a certain point in time, that is, the time when the survey is conducted. This indicates that safety climate is dynamic and may change over time, thus it is important to measure safety climate regularly to identify trends and problematic areas that need to be addressed. Third, there are likely to be many work environments within one organisation and they may have different levels of safety climate. For example, one construction organisation may be managing many projects concurrently where each project is a unique work environment with a different safety climate condition. A study by Sunindijo and Zou (2013) shows that even within one project there may be different and separate levels of safety climate. Project personnel at the managerial level may have different safety perceptions and attitudes to that of the supervisors or workers. Recognising these differences is important for aligning safety performance throughout the organisation and project.

Measurement of safety climate

Research has proposed various dimensions that should be assessed to determine safety climate level. Although these studies are good for enriching our understanding of safety climate, they may also cause confusion when deciding which dimensions should be used to measure safety climate. However, detailed investigation reveals that when many of the existing dimensions are relabelled, their number can be significantly reduced. Dimensions that are frequently used also indicate their importance. In general, the following dimensions should be included in a safety climate survey (Zou & Sunindijo, 2010):

1. *Management commitment.* This dimension implies that top-level managers should consider safety as important as the other aspects in the organisation, such as production and profit. There is also a need for top-level managers to respond decisively when a safety issue is raised. Furthermore, it is crucial for top-level managers to encourage all employees to follow safety procedures and implement initiatives to improve their safety performance. At the lower management levels, supervisors and line managers must support safety implementation through their talks and actions. They need to demonstrate their safety commitment by promoting a safe place to work and creating supportive work relationships to tackle safety issues. It is also

essential for them to include safety as an important indicator in employees' performance evaluation. Finally, this management commitment should be reflected in the authority given to safety personnel by which they must be able to enforce safety regulations and procedures at all levels in the organisation.

2. *Employee's involvement.* Safety must become one of the priorities of every employee at work. Employees need to be involved and be accountable in creating a safe workplace and improving safety performance.
3. *Safety policy, rules and procedures.* Safety policy, rules and procedures must be practical, realistic and appropriate, while also being accessible to all employees. They must meet or exceed the regulations set by the government.
4. *Safety Training.* Safety induction and training must be provided to all employees before they start their work. It is crucial for this training to be effective in providing sufficient knowledge for employees to identify safety risks and perform their work safely. Ongoing safety training should also be provided to managerial personnel and workers. The training content must be relevant to the trainees to enhance their safety learning experience and knowledge development. Chapter 5 discusses safety learning and training in more detail.

Based on these four dimensions, a sample safety climate questionnaire is given in Table 3.1. The questionnaire consists of 20 items and has been validated in the Australian construction industry context (Sunindijo & Zou, 2012).

Measuring safety climate is a practical way to assess safety performance and has been used for more than two decades. Although safety climate is relatively new in the construction industry, studies have regularly shown that safety climate is a good indicator to assess safety performance in the construction industry (Dedobbeleer & Béland, 1991; Hon & Chan, 2009; Mohamed, 2002; Siu et al., 2004; Zhou et al., 2011). The advantages of using safety climate as an indicator of safety performance include the following (Davies et al., 2001; Seo et al., 2004):

- Traditional safety indicators, such as frequency and incidence rates, are not sensitive enough to provide useful information about safety problems because they measure past performance. Construction organisations who report their incidents regularly are also disadvantaged when other organisations do not report their incidents truthfully. In comparison, safety climate can identify safety problems before they manifest themselves into accidents. In other words, it is a leading indicator, which is a measurable factor that predicts future performance instead of a lagging indicator, which is a measurable factor that shows performance of the past, that is, measuring performance based on what has transpired.
- By measuring different safety climate dimensions, construction organisations can identify problematic areas, thus providing specific areas for safety-related improvements.

Table 3.1   Safety climate measurement instrument

Dimension	No	Item	Strongly disagree	Disagree	Neither agree or disagree	Agree	Strongly agree
Management commitment	1	Top-level managers consider safety equally important as production and profits.					
	2	Top-level managers act decisively when a safety concern or problem is raised.					
	3	Top-level managers evaluate employees' safety performance and give rewards/discipline.					
	4	Top-level managers require each manager/department to improve/maintain safety performance.					
	5	Supervisors follow safety procedures in every situation, e.g. during deadline, behind schedule, planning stage.					
	6	Supervisors are committed and show interest towards safety.					
	7	Supervisors consider employees' safety performance.					
	8	Project sites are safe for employees to work.					
	9	Supportive working relationships exist in the project when it comes to safety.					
	10	The company encourages and acts upon feedback from employees on safety issues.					
	11	The company frequently holds safety campaigns or safety awareness programmes.					
	12	Safety personnel have sufficient power and authority.					
	13	Safety rules and procedures are enforced.					
Employee's involvement	14	Safety is one of the priorities when employees do their job.					
	15	Employees are involved in improving safety performance.					

(*continued overleaf*)

Table 3.1 *(continued)*

Dimension	No	Item	Strongly disagree	Disagree	Neither agree or disagree	Agree	Strongly agree
Safety policy, rules and procedures	16	The company's safety policy and safety-related information are available to all employees.					
	17	Safety rules and procedures are practical, realistic and appropriate.					
	18	It is easy to access safety-related information in the company when required.					
Safety training	19	The company's safety training programmes are effective for employees to perform their job safely.					
	20	The company's safety training programmes provide sufficient knowledge for employees to identify safety risks and hazards.					

- A safety climate survey is a valuable tool to identify trends in safety performance and to establish internal and external benchmarks.
- A safety climate survey is economical and easy to be administered.
- A safety climate survey involves all employees in the process. The process is anonymous, thus creating a sense of assurance that employees will not be identified, and encouraging them to express their true feelings without any fear of reprisal.
- Research has found a positive association between safety climate and safe work behaviour (Mohamed, 2002). One finding shows that safety climate is correlated with the safety level of the work environment, better safety practice and lower accident rate (Varonen & Mattila, 2000). Safety climate is also a predictor of injury severity and frequency (Johnson, 2007) as well as being related to self-reporting of compliance with safety procedures and participation in safety-related activities within the workplace (Neal et al., 2000).

Safety culture

Definitions of safety culture

The term safety culture can be traced back to the Chernobyl nuclear accident in 1986. At that time, a poor safety culture was identified as a contributing factor to the disaster (IAEA, 1986). Since then, the popularity of safety culture has increased and its poor implementation has been highlighted as the key source of major accidents (Cox & Flin, 1998; Health and Safety Executive, 2005). Safety culture has been defined in a variety of ways. The Confederation of British Industry (1990) defined safety culture as the ideas and beliefs that all members of the organisation share about risks, accidents and ill health. The Advisory Committee on Safety in Nuclear Installations (ACSNI) defined safety culture as the product of individual and group values, attitudes, perceptions, competencies and patterns of behaviour that determine the commitment to, and the style and proficiency of, an organisation's health and safety management (Health and Safety Commission, 1993).

Wiegmann et al. (2002, p. 8) examined various definitions of safety culture and, based on their commonalities, formulated a global definition of safety culture as follows: ' ... the enduring value and priority placed on worker and public safety by everyone in every group at every level of an organisation. It refers to the extent to which individuals and groups will commit to personal responsibility for safety, act to preserve, enhance and communicate safety concerns, strive to actively learn, adapt and modify (both individual and organisational) behaviour based on lessons learnt from mistakes, and be rewarded in a manner consistent with these values'. Fernández-Muñiz et al. (2007, p. 628) recognised the social and technical aspects of safety culture by defining it as ' ... a set of values, perceptions, attitudes and patterns of behaviour with regard to safety shared by members of the organisation; as well as a set of policies, practices and procedures relating to the reduction of employees' exposure to occupational risks, implemented at every level of the organisation, and reflecting a high level of concern and commitment to the prevention of accidents and illnesses'.

Construction safety culture can thus be defined as *an assembly of individual and group beliefs, norms, attitudes and technical practices that are concerned with minimising safety risks and exposure of workers and the public to unsafe acts and conditions in the construction environment* (Zou, 2011). Some examples of good safety culture characteristics or signs are as follows (Ostrom et al., 1993; Zou, 2011):

- The value of and belief in occupational safety are deeply and widely shared within the organisation;
- Workers have particular patterns of attitudes and beliefs regarding safety practices;
- Workers might be alert for unexpected changes and ask for help when they encounter an unfamiliar hazard;
- Workers seek and use available information that improves safety performance;
- The organisation has a safety management system in place, and this system is applied in practice and reviewed regularly;

- The organisation encourages and rewards individuals who pay attention to safety problems and who are innovative in finding ways to locate and assess hazards; and
- The organisation has systematic mechanisms to gather safety-related information, measure safety performance and bring people together to learn how to work more safely.

Dimensions of safety culture

There are three distinct but interrelated dimensions of safety culture: corporate, psychological and behavioural. The corporate dimension can be described as what the organisation has, which is reflected in the organisation's policies, operating procedures, management systems, control systems, communication flows and workflow systems. The psychological dimension is about how people feel and think about safety and safety management systems. The psychological dimension of safety culture actually refers to the safety climate of the organisation, which encompasses the attitudes and perceptions of individuals and groups towards safety. This shows that safety climate is in fact part of safety culture, a conceptualisation that has been argued in previous studies (Cooper, 2000; Cox & Flin, 1998; Glendon & Stanton, 2000; Guldenmund, 2000; Loughborough University, 2009; Wiegmann et al., 2002). The behavioural dimension is concerned with what people do within the organisation, which includes safety-related activities, actions, and behaviours exhibited by employees (Health and Safety Executive, 2005).

Although there are other studies which propose other dimensions of safety culture, Cooper (2000) explained in detail the advantages of this model. First, the corporate, psychological and behavioural dimensions of safety culture emulate accident causation relationship models. Therefore, the interrelationship among the dimensions is applicable to the accident causation chain at all levels in an organisation. The same interrelationship is in fact found in other cultural change initiatives. Second, the interrelationship among the dimensions is dynamic, thus it is suitable to portray the dynamic interaction of human and organisational systems. Third, it shows the multifaceted nature of safety culture while also making the concept tangible so that it can be systematically examined and measured within specific contexts. Fourth, the model provides organisations with a common frame of reference for the development of benchmarking partnerships with other business units or organisations.

Safety subcultures

Hopkins (2005) and Reason (1997, 2000) explained that there are five subcultures that precede safety culture as follows:

1. Informed culture

   This is a cognitive element in organisations, which relates to being alert to the possibility of unpleasant events and having the collective mindset necessary to detect, understand and recover those events before they bring about bad consequences (Reason, 2000). Organisations with an informed culture strive for system reforms instead of applying local repairs. They recognise that failures can be caused by a wide variety of unknown factors, thus they are alert for novel ways where failures and latent conditions can combine to breach the system's defences. This high level of alertness causes these organisations to always be 'preoccupied' with the possibility of failure, thus allowing them to optimally cope with the unanticipated, which is a critical component of organisational resilience. Informed culture has also been described as collective mindfulness (Weick et al., 1999).

2. Reporting culture

Reporting culture is the prerequisite of informed culture and perhaps also the most important subculture for developing safety culture. Employees at all levels must be ready and willing to report mistakes, near misses, unsafe conditions, wrong procedures and other safety concerns. This is not only about the existence of a reporting system in the organisation, but it is also about whether the occurrences are actually reported in practice. Developing this reporting culture requires employees to be proactive towards safety by always being on the lookout for things that need to be reported. They also need to have necessary skills and resources to identify and monitor things that can go wrong (Hopkins, 2005).

3. Just culture

Just culture determines the effectiveness of reporting culture. Just culture acknowledges that human beings are fallible, that is, they tend to make mistakes. Therefore, according to just culture, a risk-taking assessment should not go beyond what can be reasonably expected from these fallible human beings (Pepe & Cataldo, 2011). However, there is also a boundary which is agreed by all that some actions are unacceptable and deserve some retribution (Hudson, 2003). Organisations with just culture are willing to expose areas of weakness to improve their performance. Employees are encouraged to speak about safety issues related to their own actions and those of others. Just culture causes employees to realise that they are accountable for their actions, but will not be blamed for system faults in their work environment beyond their control (Frankel et al., 2006). Imagine an organisation that always manages mistakes with blame and punishment. Soon reports will cease as employees hide or down-play any incident. This reduces the resilience of the organisation and increases the probability of serious accidents. Blame should only be reserved for behaviour involving defiance, recklessness or malice.

4. Learning culture

Organisations should learn from safety reports or any other information conscientiously and make changes (including major reforms) as necessary to remedy or improve the situation. Learning culture also requires effective safety training programmes to develop the skills of employees to manage safety. Intrinsically, employees should have self-motivation in learning and developing their safety skills. External motivation alone, that is, motivation to learn induced by the organisation, has a limited impact on developing learning culture.

5. Flexible culture

Flexible culture means that decision-making processes are varied, depending on the urgency and the expertise of people involved. Organisations with flexible culture adapt rapidly to changes in circumstances and are able to reconfigure in the face of high-tempo operations or certain kinds of danger. In many cases, flexible culture involves shifting from the conventional hierarchy mode to a flatter structure in which control passes to task experts on the spot, and then reverts to the traditional mode once the emergency has passed. Such flexibility requires respect for the skills of the workforce, particularly the first line supervisors (Reason, 1997, 2000).

Construction organisations should focus on developing a robust safety culture that includes the five subcultures discussed above. In the following section we discuss how to measure and improve safety culture.

Safety culture maturity measurement criteria and frameworks

In 2000, a working group on Human Factors from the International Association of Oil and Gas Producers (OGP) met with academics to conduct an OGP culture study. The outcome of the study is a maturity model of safety culture which consists of five levels, namely pathological, reactive, calculative, proactive and generative. This model has been tested and appears to be robust and reliable. It specifies a pathway from less to more advanced safety culture maturity levels (Hudson, 2007). In the OGP culture study, the maturity levels from low to high are listed as follows (Hudson, 2003, 2007):

1. *Pathological*. Safety is a problem caused by workers. The main drivers of safety are the business and a desire not to get caught by the regulator.
2. *Reactive*. Organisations start to take safety seriously but there is only action after incidents.
3. *Calculative*. Safety is driven by management systems, with much collection of data. Safety is still primarily driven by management and imposed upon rather than looked for by the workforce.
4. *Proactive*. The workforce assesses and responds to safety risks proactively by using standard methods, processes and safety management system.
5. *Generative*. Safety is perceived to be an inherent part of the business. Organisations are characterised by chronic unease as a counter to complacency. There is active participation at all levels.

Another model is the Safety Culture Maturity® Model (SCMM) developed by the Keil Centre (2011) as part of a project sponsored by the UK offshore oil industry and the Health and Safety Executive (HSE). The SCMM aims to assist organisations in establishing their current level of safety culture and identifying the actions required to improve their safety culture. Although originally developed in the context of the UK offshore oil industry, the SCMM has been used successfully in many countries and sectors including aviation, road and rail transport, steelmaking, food manufacture, electronics and health care (Keil Centre, 2011). The SCMM has five levels and it is advisable for an organisation not to skip a level as it needs to progress sequentially by building on strengths and removing weaknesses of the previous level. The five maturity levels are (Fleming, 2001):

1. *Level one (emerging)*. Safety is defined in terms of technical and procedural solutions and compliance with regulations. Safety is not seen as a key business risk and the safety department is perceived to have primary responsibility for safety. Many accidents are seen as unavoidable and as part of the job. Most frontline employees are uninterested in safety.
2. *Level two (managing)*. Safety is solely defined in terms of adherence to rules and procedures and engineering controls. Accidents are seen as preventable and the majority of accidents as solely caused by the unsafe behaviour of frontline employees. Safety performance is measured in terms of lagging indicators. Senior managers are reactive in their involvement in health and safety.
3. *Level three (involving)*. The organisation is convinced that the involvement of frontline employees in safety is critical. Managers recognise that a wide range of factors cause accidents and the root causes often originate from management decisions. The majority of employees accept personal responsibility for their own safety. Safety performance is actively monitored and the data collected are used effectively.
4. *Level four (cooperating)*. The majority of employees in the organisation are convinced that safety is important from both a moral and economic point of view. Managers and frontline employees recognise that a wide range of factors cause accidents and the root causes are likely to come back to management decisions. Frontline employees accept personal responsibility for their own and fellow workers' safety. The importance of all employees feeling valued and treated fairly is recognised. The organisation puts significant effort into proactive measures to prevent accidents. Safety performance is actively monitored using all data available.
5. *Level five (continually improving)*. The prevention of all injuries or harm to employees is a core company value. There is no feeling of complacency as people live with the paranoia that their next accident is just round the corner. The organisation uses a range of indicators to monitor performance but it is not performance-driven, as it has confidence in its safety processes. The organisation is constantly striving to be better and find better ways of improving hazard control mechanisms. All employees share the belief that safety is a critical aspect of their job and accept that the prevention of non-work injuries is important.

Based on the above discussions, we adopted the SCMM developed by the Keil Centre for the construction industry. We felt that this model provides more accuracy due to its less threatening nature and its use of everyday words. Furthermore, the SCMM has been applied successfully in other industries. Concerning the issue with the label/name of each maturity level, we argued that it would be unproductive to focus on semantics and miss the important intention to assist construction organisations to measure their safety culture maturity levels. We developed a safety culture framework as depicted in Figure 3.1 to integrate all the aspects discussed in the previous sections.

The five subcultures are just, reporting, informed, flexible, and learning cultures. As discussed previously, these subcultures are the underlying aspects of

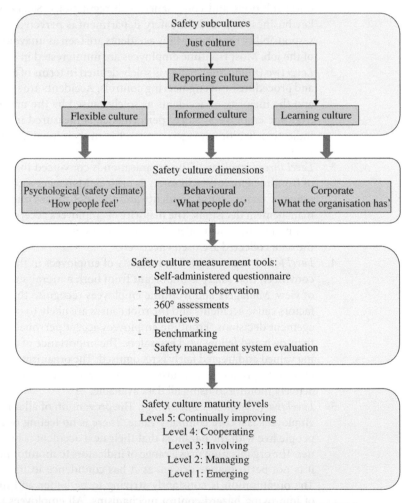

Figure 3.1   Construction safety culture maturity measurement framework

safety culture. They essentially determine the maturity of safety culture in an organisation. These subcultures are manifested in the three dimensions of safety culture: what people feel (psychological), what people do (behavioural) and what the organisation has (corporate). Because of the manifestation of safety culture through its three dimensions, we are able to observe and assess safety culture. Due to its multi-dimensional and multi-characteristics nature, obtaining a complete picture of safety culture requires a range of tools, such as:

- Self-administered questionnaires to measure the perceptions of employees concerning certain aspects of safety in the organisation. A safety climate questionnaire presented in Table 3.1 is an example of this.
- Behavioural observations to assess employees' safety behaviour by observing how they really behave in the workplace.
- 360° assessments to verify the results of self-administered questionnaires and behavioural observations. Respondents may answer questionnaires or behave in certain ways when they are conscious that they are being researched upon, which may not reflect the 'normal' condition. 360° assessments can be used to reduce this kind of bias.
- Interviews to gain deeper insight concerning phenomena that are being assessed.
- Benchmarking to compare safety performance across business units or to compare the organisation's safety performance against the 'best practice'.
- Safety management system evaluation, monitoring and auditing to assess the robustness and effectiveness of the system, which is concerned with the corporate dimension of safety culture.

These tools allow construction organisations to obtain an overall picture of their safety culture, thus helping them determine their safety culture maturity level, whether it is at the emerging, managing, involving, cooperating or continually improving level. Finally, this understanding of safety culture will enable those organisations to strategise in improving their safety performance.

Safety culture maturity measurement instrument

This section focuses on defining the measurement criteria of each safety culture maturity level in order to assist construction organisations to measure the dimensions of safety culture underpinned by the five subcultures. The method for developing the measurement criteria is based on the works done by Parker et al. (2006) and Filho et al. (2010) in which they conducted an exploratory study to generate a theory-based framework that could be used by organisations in the oil and gas industry to understand their safety culture. We adapted their works and other safety culture studies, including Fleming (2001), Frankel et al. (2006), Health and Safety Executive (2005), Hudson (2003), Lardner (2002), Pepe and Cataldo (2011), Reason (1998), and Wiegmann et al. (2002), to the context of construction and engineering industry while also integrating the five subcultures and safety culture dimensions into the measurement criteria. As stated earlier, safety culture is complex and multi-dimensional, thus it is challenging to gauge the level of safety culture because it is beyond the scope of any single method (Glendon & Stanton, 2000; Lee & Harrison, 2000). Therefore, the measurement criteria proposed here, although convenient, are limited in terms of details and should only be considered as general guidelines to give a basic idea about the safety culture maturity level of an organisation.

Psychological dimension

The psychological dimension of safety culture is about the attitudes and perceptions of people towards safety. The characteristics of the psychological dimension in each safety culture maturity level are described as follows:

- Level 1 – emerging: Employees perceive that safety is fundamentally a waste of time and is necessary only to comply with regulations.
- Level 2 – managing: Employees perceive that safety is about following rules and procedures, while accidents happen because these rules and regulations are not followed, typically by frontline employees.
- Level 3 – involving: Employees perceive that every person is responsible for his or her own safety, while managers believe that accidents are generally caused by poor management decisions.
- Level 4 – cooperating: Employees perceive that they are responsible for their own and other people's safety and there is a belief throughout the organisation that poor management decisions are the root causes of accidents.
- Level 5 – continually improving: Employees perceive that safety is a critical aspect of their job and an indicator of performance, thus it should be continually improved.

Table 3.2 shows examples of safety attitudes and perceptions in each maturity level which are also classified further based on the five safety subcultures.

Behavioural dimension

The behavioural dimension of safety culture is concerned with what employees do within the organisation in terms of safety. The characteristics of the behavioural dimension in each safety culture maturity level are described as follows:

- Level 1 – emerging: Employees do not consider safety at work because they focus on the other project objectives, such as cost and time.
- Level 2 – managing: Employees behave safely when they are monitored by safety personnel.

Table 3.2 The psychological dimension of safety culture and its maturity levels

Subculture	Level 1 – emerging	Level 2 – managing	Level 3 – involving	Level 4 – cooperating	Level 5 – continually improving
Just	- Accidents are bad luck - Some people are prone to get injured	- Accidents happen because workers do not follow safety procedures - Workers got injured because they did not wear PPE	Workers involved in an accident and their managers are responsible for the accident	The team contributes to and is responsible for an accident	The organisation and its system contribute to and are responsible for an accident
Reporting	- Safety reporting is a waste of time and resources - Safety issues must be concealed	Safety reporting is required according to rules and regulations - Safety issues should be reported when asked by managers	- Managers are the ones responsible for reporting safety issues - Employees can report safety issues to their managers	Employees can report safety issues through an established system freely	Employees can report safety issues and offer suggestions through an established system freely
Informed	- Accidents are part of the job - Construction is a dangerous industry	- Safety is the responsibility of safety personnel - Employees should comply with safety rules and regulations	- Every employee is responsible for his or her own safety - Managers should control their employees	Employees and managers are responsible for their own and team members' safety	Everyone in the organisation is involved and responsible for managing and improving safety

Psychological

(continued overleaf)

Table 3.2 (*continued*)

Subculture	Level 1 – emerging	Level 2 – managing	Level 3 – involving	Level 4 – cooperating	Level 5 – continually improving
Flexible	Safety is non-existent	Safety personnel should make safety-related decisions	Safety personnel and managers should make safety-related decisions	Besides safety personnel and managers, some employees are empowered to make safety-related decisions	Everyone in the organisation is empowered to make safety-related decisions
Learning	Safety training is a waste of time and resources	Safety training is required	- Safety training and learning are necessary for 'my own safety' - Safety learning is motivated by extrinsic factors	Safety training and learning are necessary for 'my own and other people's safety Safety learning is motivated by extrinsic and intrinsic factors	Safety learning is motivated by intrinsic and extrinsic factors

- Level 3 – involving: Employees behave safely for their personal gain.
- Level 4 – cooperating: Employees behave safely and remind others to work safely.
- Level 5 – continually improving: Employees integrate safety into daily activities and actively find ways to improve safety performance.

Table 3.3 shows examples of safety behaviour in each maturity level which are also classified further based on the five safety subcultures.

Corporate dimension

The corporate dimension of safety culture is concerned with what policy and management system the organisation has in place, in relation to safety. The characteristics of the corporate dimension in each safety culture maturity level are described as follows:

- Level 1 – emerging: There is no formal safety management system in the organisation.
- Level 2 – managing: A formal safety management system is established by safety personnel and is followed by employees mechanically.
- Level 3 – involving: Safety personnel and managers lead the implementation of the safety management system. They recognise the effectiveness of the system to manage safety performance. Employees are involved in the process.
- Level 4 – cooperating: Not only the managers and safety personnel, but also the employees recognise the effectiveness of the system to manage safety and use the system in performing their daily activities.
- Level 5 – continually improving: The safety management system is audited periodically by external parties to improve its overall effectiveness.

Table 3.4 shows examples of safety management system development and implementation in each maturity level, which are also classified further, on the basis of the five safety subcultures.

Case studies

Case 1 Fluor's 'zero incident' safety programme

Fluor Corporation was founded in 1912 in the USA and currently is one of the largest construction organisations in the world. Fluor employs more than 41,000 employees in 79 countries (Fluor Corporation, 2013b). Fluor recognises that successfully managing health, safety and environmental (HSE) issues is part of its business strategy. The organisation strives to identify HSE risks arising from its activities and reduce them to the lowest practical level to

Table 3.3 The behavioural dimension of safety culture and its maturity levels

Subculture	Level 1 – emerging	Level 2 – managing	Level 3 – involving	Level 4 – cooperating	Level 5 – Continually improving
Just	Employees look after themselves and blame others after an accident	- Employees attempt to cover up accidents - Accident-prone employees are removed	- Managers are upset by an accident due to its impact on 'statistics' - Employees participate in accident investigation and safety audit as required	- Managers care about the well-being of those involved in an accident. - Employees are cooperative during investigation or safety audit	Everyone in the organisation collaborates in implementing safety improvement measures
Reporting	Employees do not report safety issues	Safety personnel prepare a report especially after an accident as directed	- Employees prepare a periodic safety report - Employees collect safety statistics according to procedures.	- Employees proactively communicate safety issues - Managers encourage employees to report safety issues - Managers share safety issues with all employees	Everyone in the organisation actively reports safety issues and uses safety information to improve safety performance

Behavioural

Informed	Employees work as quickly and as economically as possible	- Safety personnel monitor safety performance - Employees behave safely when monitored	- Employees follow safety procedures as required - Periodic inspections by managers to ensure compliance	- Employees remind one another about safety - Managers consider safety a priority in making decisions	Everyone in the organisation embraces safety as part of his/her job and actively tells each other about hazards while offering ideas to improve safety performance
Flexible	Employees use 'commonsense' to finish the work on time and on budget	Safety personnel tells others to work safely	Safety personnel and managers tell employees to work safely and to think safety	Safety personnel, managers, lead safety implementation and employees cooperate in the training process	Everyone in the organisation is able to make safety-related decisions relevant to them
Learning	Employees participate in safety training so they are allowed to work	- Employees participate in safety training as required - Safety personnel provide input on safety training programme design and implementation	- Managers encourage employees to improve safety knowledge - Employees learn about safety because of intrinsic motivators	Employees and managers learn about safety because of intrinsic and extrinsic motivators	Everyone in the organisation learns about safety continually mainly due to intrinsic motivators

Table 3.4 The corporate dimension of safety culture and its maturity levels

	Subculture	Level 1 – emerging	Level 2 – managing	Level 3 – involving	Level 4 – cooperating	Level 5 – continually improving
Corporate	Just	There are consequences for employees involved in an accident	Poor safety performance leads to disincentive, while bonuses are tied to lost time injury (LTI)	- Good safety performance is acknolwdged by lip service or safety awards - Incidence rate is used to calculate bonuses	- Safety evaluation is based on process rather than outcome - Good performance leads to rewards and is considered in promotion reviews	A whole- system approach including the interaction between systems and employees is observed to evaluate and improve the safety management system
	Reporting	- There is no reporting system - Investigation takes place only after a serious accident to meet legal requirements	- Reporting exists but is focused on finding the guilty parties - There is little follow- up to track remedial actions	- There are fixed procedures and requirements for reporting accidents - Investigation focuses on localised or workplace issues	- Reports offer insight on 'why' rather than 'what' of accidents - Investigation has systematic follow-up - Reports are distributed company-wide to share lessons learnt	- Reports are integrated to identify root causes and improve performance - Follow- up and feedback mechanisms are systematic to ensure that changes occur and are maintained
	Informed	- There is no formal safety management system - On-site work plan focuses on the most economical and fastest methods, and does not consider safety.	- There is no formal follow- up system after an accident - Safety personnel make safety calls to comply with regulations and employees follow them passively	- Managers use safety management system to consider safety when making decisions - Job safety techniques and training are introduced to all employees	- Managers and employees consider the safety management system as effective and relevant to their work - Safety is seen as a priority when making decision by management and employees	- Job safety techniques are revised regularly in a defined process to achieve best practice - Formal and informal monitoring of safety issues is done by all to anticipate problems

Flexible	No system to implement safety and measure performance	- Safety personnel are appointed to solve safety problems reactively - Minimal follow-up	There is a clear structure showing safety responsibilities of safety personnel and managers	- Safety personnel are respected by employees - Managers lead safety implementation and employees coorperate	Safety responsibilities are distributed throughout the company because employees are knowledgeable and skilful in safety implementation
Learning	Provide safety training as required by law	Additional (voluntary) safety training is provided after an accident	Safety management system includes clear safety training procedures for all employees	Managers and employees participate in safety training programme development, and monitor the effectiveness of safety training to find ways to improve	- Everyone in the organisation proactively learns about safety and shares best practices with others - Managers collaborate with external parties to improve safety performance - Employees constantly relfect on safety to find ways for improvement.

protect the environment and the well-being of its stakeholders, specifically, its employees, clients, shareholders, subcontractors and communities (Fluor Corporation, 2013c). Fluor's commitment is motivated by the belief that HSE issues can be systematically identified and managed, that no job is worth a loss of life or injury, and that 'zero incidents' is an attainable objective (Fluor Corporation, 2014).

Safety programme implementation

To achieve its 'zero incidents' goals, Fluor has implemented a number of strategies, which cover management approach, engagement and training programmes, and HSE management system. Fluor's management is committed to its HSE policy. The HSE commitment is demonstrated in the following examples (Fluor Corporation, 2013a):

- Opening each meeting with an HSE topic for discussion;
- Managing HSE activities in the same manner as productivity, quality and scheduling;
- Integrating HSE into the organisation's functions and work processes;
- Accepting accountability for HSE activities;
- Championing HSE practices across all operations;
- Employing 850 dedicated HSE professionals around the world.

Fluor promotes both organisational and personal accountabilities for HSE performance. For site managers, Fluor was one of the first contractors to award monetary incentives for superior HSE performance. Fluor implements a variety of programmes to recognise desired HSE behaviours and results for groups and individuals. Senior management celebrates significant achievements and communicates these across the organisation. Specific groups within the organisation tie HSE performance to merit increases, bonuses, and other incentive compensation programmes. Fluor also involves employees in the design of recognition and reward systems.

Fluor engages its stakeholders on HSE-related issues as part of its regular operations. HSE-related communications are conducted regularly to reinforce Fluor's HSE mindset, which calls all employees to own and address any HSE issue that may arise, even if it is not their own. Fluor also collaborates with subcontractors, clients, research organisations, professional bodies and trade unions to promote continual improvement in HSE matters.

It is estimated that more than 40% of accidents occur during the first year of working in the industry (Root, 1981). Fluor provides HSE orientation and preventive training to every new employee. In addition to orientation training, employees participate in comprehensive training programmes in the areas of compliance, behaviour and culture. The safety leadership training module for managers emphasises employee coaching, communications, behaviour modification and team building skills.

Fluor's HSE management system, known as 'Zero Incidents' programme, is designed to integrate the management of HSE practices into one programme that embraces a project or activity from beginning to end and is consistent throughout the company globally. The system is comparable to ISO 14001, OHSAS 18001, ANSI Z-10, and US OSHA Voluntary Protection Program, thus integrating the highest international standards into each project phase to ensure consistent performance. A key component of the system is the audit programme, which emphasises the use of leading indicators in HSE programme development and coordination; management-in-action; training, communication and HSE culture initiatives and field execution.

At the project level, Fluor's proprietary Managing to Zero (MS_20) programme measures HSE indicators to aid improvement of working conditions on a continual basis. MS_20 is a web-based, centralised database programme for tracking and trending leading and lagging indicators. Information obtained from key indicators, such as daily audits, near-miss incidents and hazards eliminated, are collected to produce trend results. Such a system enables Fluor's HSE professionals to share collective lessons learnt and best practices and to collaborate on real-time issues across time zones. Project management analyses these trend results to prevent or reduce incidents and address topics in the areas of root cause analysis, implementation of new regulatory standards and emerging trends.

Key lessons

The Fluor case study reveals the following lessons learnt for successful HSE programmes:

- Top management is committed to the safety objectives and goals, management and implementation processes, audit and feedback mechanisms and regular review and improvement of the HSE programme. Management takes a proactive approach to create safe work environments and is accountable for providing safety education and training for all employees, continuously reviewing the programme to identify potential areas of improvement and ensuring a thorough evaluation of all incidents.
- The organisation works closely with clients, suppliers and subcontractors to ensure that the HSE programme is comprehensive.
- The corporate safety culture includes a belief that all incidents and injuries are preventable and that the objective of 'zero incidents' is achievable through education and training, monitoring, audit and feedback mechanisms and continuous improvement.
- The organisation has standard procedures for reporting HSE performance, for investigating and recording all incidents and complaints and for taking appropriate corrective action to avoid recurrence.
- Records are maintained and information and statistics are reported to corporate and regional HSE management. Formal auditing procedures

are defined and implemented. Deficiencies identified during audits are recorded formally, their implications are assessed and corrective actions are prioritised.

- Excellent safety performance is rewarded not only verbally but also in monetary terms.
- Knowledge gained is recorded and applied to future projects to increase the organisation's safety management capability and maturity.

Case 2 gammon's 'safety first, zero accident' programme

Gammon Construction originated as a construction business in India in 1919. Soon after, several branches were established in Asia, the Middle East and Africa. Currently it is a leading construction and engineering contractor headquartered in Hong Kong. It employs more than 8000 people and has a strong presence in the Southeast Asia region (Gammon Construction, 2012a).

To improve its safety performance, Gammon carried out a large-scale survey to examine the safety culture of workers, including safety-related values, attitudes and behaviour, on 50 projects. The results showed that workers' safety behaviours were dependent on various factors, such as project nature, client requirements, supervisors' attitudes and management support and commitment. Following this study, Gammon's management initiated a programme called 'Safety First, Zero Accident' aimed at improving workplace safety by changing workers' behaviours (Zou, 2011).

Today Gammon focuses on eight aspects to develop its safety culture: mindful, learning, informed, fair and respectful, design and engineering, plant and equipment, process and people. The latest survey shows a significant improvement in all aspects. The organisation has also succeeded in bringing down its overall accident incident rate, which currently stands far below the industry average. Over the years, Gammon realised that preventing accidents by modifying worker behaviour only had limited results. Recently, it reinvigorated its safety focus on the elimination of risks based on the layers of protection principle (Gammon Construction, 2013).

Safety programme implementation

In implementing its safety programme, the organisation focuses on recognising the importance of management leadership and commitment, safety risk assessment at the design stage, stakeholder engagement, ongoing safety training and the application of information technology. These topics are discussed in the following paragraphs.

Management leadership and commitment to safety are the most important aspects of safety implementation in Gammon, as they support the other implementation strategies. At the corporate level, safety is one of the core values. Gammon's principle is to actively challenge the construction process to reduce and remove risk. The organisation applies this principle by making safety personal and meaningful, managers accepting their responsibility for keeping people safe, making it easy to work safely, integrating safety as part of effective and efficient construction and always putting safety first in a conflict of interest (Gammon Construction, 2012a). At the work-site level, Gammon realises the importance of providing visible leadership through the engagement of the frontline supervisors who directly interface with the workers. Site walks which involve a mix of management and various disciplines are held weekly to engage project teams and find solutions together. This activity strengthens the links between the management and the frontline staff, while making practical solutions more forthcoming. When incidents occur, Gammon takes immediate action to investigate their causes and to rectify them. Directors will also hold on-site reviews of all incidents and engage the workforce to seek solutions on how to prevent similar incidents in the future (Gammon Construction, 2013).

In Gammon, safety risk assessment is conducted at the design stage to iden-tify possible hazards during project implementation, for example, construction and operation, and to find possible solutions. The design process has a signifi-cant impact on project implementation and design changes are regularly made to reduce hazards.

Gammon realises that stakeholder engagement is critical to ensure effective safety implementation. One strategy to engage stakeholders is the Gammon Annual Safety Conference. The purpose of the conference is to promote safety culture in the construction industry and demonstrate the organisation's commitment towards its staff and the public. The conferences have attracted more than 550 participants every year, including clients, subcontractors, suppliers, and internal staff. A number of guests are invited to speak and share their insights and experiences on safety issues through speeches and discussion forums (Gammon Construction, 2012b).

To maintain consistency in health and safety standards among subcontrac-tors and suppliers, Gammon has a full-day training course entitled 'HSE Man-agement System for Subcontractors' that provides essential safety and health training. Since January 2010, Gammon's Zero Harm Induction Centre has been training Gammon staff and subcontractors on the elements of Zero Harm. All Gammon monthly and daily-paid staff and workers, including subcontractors at all levels, are required to take part in the half-day training programme, which has been specifically designed to focus on removing the five risks most com-monly associated with fatalities, including working at height, falling objects, electrical equipment, moving plant and people, and drowning. The ultimate objective of this training programme is to provide participants with health and safety knowledge specific to Gammon's work environments. Apart from rais-ing health and safety awareness, the programme also aims to change workers' behaviour and ensure individuals' well-being by creating an accident-free work-place to the benefit of every party (Gammon Construction, 2012b).

In applying information technology to improve safety performance, Gam-mon has used enterprise resource planning software to provide a web-based system for reporting all environmental, health, safety and security incidents. It provides comprehensive data analysis for identifying trends or patterns, docu-menting accident cause investigations and presenting HSE data.

Key lessons

The following elements of the Gammon case study provide lessons learnt for successful safety programmes:

- The organisation concentrates its promotion of safety leadership and commitment on two different levels: workers and corporate management.
- Regular engagements between the management and frontline staff improve communication and provide practical safety solutions.
- Safety risk assessment at the design phase reduces and mitigates safety risks.
- Subcontractors and suppliers receive training in standardised performance across the supply chain. Other stakeholder engagement strategies are also applied to improve safety culture in the industry.

Case 3 John Holland's 'no harm' safety programme

John Holland was established in 1949 and is one of the largest construction organisations in Australia, with more than 6500 employees nationwide (John Holland, 2014). 'No Harm' is a belief that in any circumstance harm and damage can be prevented, and that everything practicable must be done to get it right the first time. John Holland's 'No Harm' focuses on behaviours at all levels because they can be defined, observed and measured. The organisation provides better plant, facilities and business processes to make 'No Harm' a reality. However, it recognises that excellent performance also depends on people demonstrating behaviours that make the systems work in practice. The right tools and methods to understand behaviour and reinforce safe behaviours are key to foster a strong safety culture (John Holland, n.d.).

Safety Programme Implementation

John Holland's Safety Strategy (2010–2013) has been developed to bring the 'No Harm' vision to life. The strategy has identified four elements fundamental to safety improvement: leadership, risk management, governance and capability. John Holland has implemented a number of key initiatives to demonstrate its safety leadership. Some of these initiatives are (John Holland, 2013, n.d.):

- 'No Harm' behavioural framework to assess personal strengths and areas for improvement.
- Comprehensive safety training and development programmes, such as the Passport to Safety Excellence Program (PSEP), 5-day Health and Safety Representative (HSR) Training Program, and Start Card process.
- Working with industry partners and the Office of the Federal Safety Commissioner on key areas to drive positive change across the industry.
- Ongoing due diligence programme to audit the safety management system and evaluate safety performance.

- Forming the Executive Safety Leadership Team to review safety performance within the business monthly.
- Global Mandatory Requirements for safety to outline the control strategies and minimum standards for the key risks that people are exposed to every day across the organisation.
- Safety Achievers Awards for employees who demonstrate the highest level of safety leadership.

John Holland's risk management approach is to align three critical areas of risk: Safety, Quality, and Environment (SQE). The focus on SQE risk management at all phases of business lifecycle provides a significant level of visibility to the risks associated with the work, and a rigour to the control strategies, much earlier in the delivery lifecycle. The result is an increased opportunity to plan and act before doing any work, and to involve the right people at the right time in the planning process.

To ensure that the 'No Harm' vision and safety management system are applied throughout the organisation, John Holland has developed a governance framework and consultative structure. The governance framework comprises a six-tiered approach that aligns with the hierarchy of the organisation. The framework facilitates effective consultation and communication between worksites, Group Health and Safety Committee, Executive Leadership Team and Board OHS Committee.

John Holland addresses the capability of its employees by employing various safety training programmes. The Start Card has been designed to raise every employee's safety awareness and to create safety conversations between supervisors and employees. The Start Card process empowers people to determine whether or not to proceed with the task or activity where they identify conditions that may pose a risk of injury, illness, property or environmental damage. The PSEP is a nationally recognised qualification, Certificate IV in Safety Leadership (OHS) – Construction, which is John Holland's national training programme. It is designed to equip people in safety critical positions with the skills and behavioural competencies required to effectively carry out their role (John Holland, n.d.).

Key lessons

Some lessons learnt from the John Holland case study are:

- A governance framework and consultative structure to ensure that safety implementation is aligned across management levels.
- Using risk management approach to align safety, quality and environment; thus safety is considered a priority in any project.
- Planning for safety early in the project lifecycle.
- Comprehensive training programmes, including a nationally recognised qualification, thus motivating employees to participate.

- Focusing on behaviours which are considered as more tangible.
- Rewards are given to employees who demonstrate excellent safety leadership.
- Engaging external stakeholders, such as the Federal Safety Commissioner, and developing a nationally recognised Certificate IV safety training programme.

Utility of safety culture

When discussing safety culture, we need to be aware of its utility and the possible misalignment across management levels within an organisation, for example, misalignment across the organisation, project, and on-site work crew levels. Research has shown that there is a real issue of misalignment in implementing safety policy from the governments' policy development to the company's boardroom decision, and to on-site implementation. Habibi and Fereidan (2009) assessed the attitudes of three levels of refinery personnel in Iran, including top management, supervisory staff and frontline workers, towards safety culture in the organisation, and they found significant differences between the management level and both the supervisory staff and frontline workers. In the Thai construction industry, Pungvongsanuraks and Chinda (2010) found a misalignment of safety culture perceptions between management and workers, where top management believes that safety empowerment and training are important whilst workers consider this a waste of time because they want to focus their effort on maintaining their productiveness. Likewise, Fung et al. (2005) found divergences of behaviour, attitude and perception towards safety culture among top management, supervisory staff and frontline workers in the Hong Kong construction industry. Similarly, Chen et al. (2012) found a gap in construction safety climate awareness between management and workers in Taiwan. This situation also happens in Australia, where Sunindijo and Zou (2013) found that there are different safety perceptions between the management and supervisory levels. Generally speaking, managers perceived higher levels of safety climate than the supervisors. This is disconcerting as it indicates that the managers are detached from the real safety conditions on-site. Measuring safety culture, therefore, should be done in a systematic and holistic manner. There is a tendency to assume that safety culture is determined on the basis of the perceptions of senior managers. There is a danger that these perceptions may not reflect the real conditions at the work-site level.

Safety culture can also be seen from 'outside the box' perspectives. Nowadays, construction organisations tend to focus only on safety culture within their own organisations. This is inadequate because of the nature of the industry, where subcontracting practice and the involvement of numerous stakeholders are common. Therefore, the challenge is to develop safety culture across the supply chain, that is, inter-organisational safety culture (Fang & Wu, 2012). To make the situation more complex, some organisations are operating regionally and globally, thus facing differing cultural backgrounds and contextual conditions, which will greatly influence the interpretation and implementation of safety policy and safety systems. For example, Chen and Jin (2013) found that the incidence rates of a contractor in the USA varied greatly across regional branches.

As a result of the issues described above, research continues to develop more complex and varied conceptualisations of safety culture to provide more detailed models of the environments they represent. However, we need to be careful here because increased complexity does not always translate to increased utility. The increased model complexity causes concerns that academic tools and systems have become impractical and unwieldy when applied to practice (Sherratt, 2014). Sherratt (2014) found that workforce engagement is the most challenging element of safety management in the construction and engineering industry. In this case, the utility of safety culture is about the ability to support the engagement and communication of safety in the organisation, on its projects, and all along its supply chains. Despite a positive safety culture and strong management commitment, without this employee engagement, accidents may still occur due to the autonomous nature of work on construction sites and the different management layers across the organisation. Research which adopts ethnographic and constructionist approaches to focus on people and their interactions, behaviours and attitudes is essential to understand complex social interactions and to provide effective engagement strategies in different contexts. Chapter 7 discusses an alternative mixed methods research design, which can assist researchers through a nexus between research and practice to achieve this goal.

Conclusions

It is necessary to foster a strong safety culture in strategic safety management. Intrinsically, safety culture is the expression of strategic safety management. Lack of safety culture maturity indicates ineffectiveness in the implementation of strategic safety management. Safety culture can be classified into three dimensions: psychological (what people feel, think and believe), behavioural (what people do) and corporate (what the organisation has). Organisations which aspire to develop safety culture should focus on five subcultures: informed culture, reporting culture, just culture, learning culture and flexible culture. Measurement criteria have been included so that construction organisations can measure their safety culture maturity on these subcultures, identify their weak areas and develop strategies for improving their safety culture.

Three case studies have been presented to demonstrate strong safety culture in practice. The case organisations recognise the importance of safety in their business, thus they are committed to improving safety. Seven common themes can be drawn from these case studies. First, all cases emphasise the importance of human factors, including attitudes, beliefs, values, mindsets and behaviour. In all cases, programmes were set up to shape the belief and value that all incidents and injuries are preventable and unacceptable to management and workers. Secondly, the cases show the importance of a commitment and leadership from top management as a fundamental factor in shaping a strong construction safety culture. Thirdly, there is a need to engage the entire supply chain and every project stakeholder, each of whom has interests in and influence over safety. Fourthly, safety risk management systems were established in all cases to support ongoing monitoring through reporting, auditing, and reviewing safety performance, as well as updating communications and procedures. Fifthly, it is important to establish clear authority and accountability for safety. Safe behaviour and good safety performance should also be rewarded. Sixthly, a safety knowledge database (typically online) is a useful tool to capture lessons learnt so that risk management principles and techniques can be integrated into safety management processes and be applied to future projects. Seventhly, it is essential to address safety early in the project lifecycle, that is, in the planning and design stages, rather than waiting until the construction stage, and to extend this consideration to the entire project lifecycle.

Ultimately, the success of a safety programme depends as much on people's attitudes and behaviours as it does on safety programme design. Although it is easy to bring about behavioural change, it is very difficult to maintain the changes achieved. Developing safety culture for a construction project or organisation does not occur overnight; it is a journey rather than a destination, and it requires a commitment from top management and supervisors, right down to individual employees' involvement over an extended period of time. Construction organisations adopting a new approach to safety management must continue to champion the new philosophy, value, and belief, and to monitor performance, while learning the lessons and feeding them back into business processes and management practices. These efforts must be supported by an effective training, motivation and performance appraisal system to reinforce appropriate behaviours. The vision of 'zero incidents' can be achieved only by balancing the two sides of the coin, the 'science' and 'art' of safety management in construction and engineering.

References

Antonsen, S. (2009). *Safety Culture: Theory, Method and Improvement*. Surrey, England: Ashgate.

Brown, R. L., & Holmes, H. (1986). The use of a factor-analytic procedure for assessing the validity of an employee safety climate model. *Accident Analysis & Prevention*, *18*(6), 455–470.

Chen, Q., & Jin, R. (2013). Safety4Site commitment to enhance jobsite safety management and performance. *Journal of Construction Engineering and Management*, *138*(4), 509–519.

Chen, W. T., Liu, S.-S., Liou, S.-W., & Sun, W. Z. (2012). Disrepancies between management and labor perceptions of construction site safety. *Applied Mechanics and Materials*, *174–177*, 2950–2956.

Choudhry, R., Fang, D., & Mohamed, S. (2007). The nature of safety culture: A survey of the state-of-the-art. *Safety Science*, *45*(10), 993–1012. doi: 10.1016/j.ssci.2006.09.003

Confederation of British Industry. (1990). *Developing a Safety Culture-Business for Safety*. London: Confederation of British Industry.

Cooper, M. D. (2000). Towards a model of safety culture. *Safety Science, 36*(2), 111–136.

Cooper, M. D., & Phillips, R. A. (2004). Exploratory analysis of the safety climate and safety behavior relationship. *Journal of Safety Research, 35*(5), 497–512.

Cox, S. J., & Flin, R. (1998). Safety culture: Philosopher's stone or man of straw? *Work & Stress, 12*(3), 189–201. doi: 10.1080/02678379808256861

Coyle, I. R., Sleeman, S. D., & Adams, N. (1995). Safety climate. *Journal of Safety Research, 26*(4), 247–254.

Davies, F., Spencer, R., & Dooley, K. (2001). *Summary Guide to Safety Climate Tools.* Norwich: HSE Books.

Dedobbeleer, N., & Béland, F. (1991). A safety climate measure for construction sites. *Journal of Safety Research, 22*(2), 97–103.

Fang, D., & Wu, H. (2012). Safety culture in construction projects. Paper presented at the CIB W099 International Conference on Modelling and Building Health and Safety, Singapore.

Fernández-Muñiz, B., Montes-Peón, J. M., & Vázquez-Ordás, C. J. (2007). Safety culture: Analysis of the causal relationships between its key dimensions. *Journal of Safety Research, 38*(6), 627–641. doi: 10.1016/j.jsr.2007.09.001

Filho, A. P. G., Andrade, J. C. S., & Marinho, M. M. d. O. (2010). A safety culture maturity model for petrochemical companies in Brazil. *Safety Science, 48*(5), 615–624. doi: 10.1016/j.ssci.2010.01.012

Fleming, M. (2001). *Safety Culture Maturity Model – Offshore Technology Report 200049.* Norwich: HSE Books.

Fluor Corporation. (2013a). *2012 Sustainability Report.* Irving: Fluor Corporation.

Fluor Corporation. (2013b). *Corporate Profile.* Irving: Fluor Corporation.

Fluor Corporation. (2013c). *Health, Safety, and Environmental Policy.* Irving: Fluor Corporation.

Fluor Corporation. (2014). Fluor's Sustainable Commitment to Health, Safety, and the Environment Retrieved 15 Jan, 2014, from http://www.fluor.com/sustainability/ health_safety_environmental/Pages/dcfault.aspx

Frankel, A. S., Leonard, M. W., & Denham, C. R. (2006). Fair and just culture, team behavior, and leadership engagement: The tools to achieve high reliability. *Health Services Research, 41*(4p2), 1690–1709. doi: 10.1111/j.1475-6773.2006.00572.x

Fung, I. W. H., Tam, C. M., Tung, K. C. F., & Man, A. S. K. (2005). Safety cultural divergences among management, supervisory and worker groups in Hong Kong construction industry. *International Journal of Project Management, 23*(7), 504–512.

Gammon Construction. (2012a). About Us Retrieved 16 Jan, 2014, from http://www .gammonconstruction.com/en/html/about-us/corporate-profile.html

Gammon Construction. (2012b). Sustainability Framework: Health and Safety Retrieved 16 Jan, 2014, from http://www.gammonconstruction.com/en/html/sustainability/ health.html

Gammon Construction. (2013). *The Power of 10, Sustainability Report 2012.* Hong Kong: Gammon Construction.

Glendon, A. I., & Stanton, N. A. (2000). Perspectives on safety culture. *Safety Science, 34*(1-3), 193–214.

Guldenmund, F. W. (2000). The nature of safety culture: A review of theory and research. *Safety Science, 34*(1–3), 215–257.

Habibi, E., & Fereidan, M. (2009). Safety cultural assessment among management, supervisory and worker groups in a tar refinery plant. *Journal of Research in Health Sciences, 9*(1), 30–36.

Hahn, S. E., & Murphy, L. R. (2008). A short scale for measuring safety climate. *Safety Science, 46*(7), 1047–1066. doi: 10.1016/j.ssci.2007.06.002

Haukelid, K. (2008). Theories of (safety) culture revisited—An anthropological approach. *Safety Science, 46*(3), 413–426. doi: 10.1016/j.ssci.2007.05.014

Health and Safety Commission. (1993). *ACSNI Study Group on Human Factors, Third Report*. London: HSE Books.

Health and Safety Executive. (2005). *A Review of Safety Culture and Safety Climate Literature for the Development of the Safety Culture Inspection Toolkit*. Bristol: HSE Books.

Hon, C., & Chan, A. (2009). Safety climate: Recent developments and future implications. Paper presented at the CIB W099 Conference, Working Together: Planning, Designing and Building a Healthy and Safe Construction Industry, Melbourne, Australia.

Hopkins, A. (2005). *Safety, Culture and Risk: The Organisational Causes of Disasters*. Sydney: CCH.

Hudson, P. (2003). Applying the lessons of high risk industries to health care. *Quality and Safety in Health Care, 12*(Suppl I), i1–i12.

Hudson, P. (2007). Implementing a safety culture in a major multi-national. *Safety Science, 45*(6), 697–722. doi: 10.1016/j.ssci.2007.04.005

IAEA. (1986). Summary Report on the Post-Accident Review Meeting on the Chernobyl Accident International Safety Advisory Group, Safety Series 75-INSAG-1 Vienna: International Atomic Energy Agency.

John Holland. (2013). *John Holland Annual Review 2012*. Abbotsford: John Holland.

John Holland. (2014). About Us Retrieved 16 Jan, 2014, from http://www.johnholland.com.au/Documents.asp?ID=13986&Title=Who+We+Are

John Holland. (n.d.). *John Holland No Harm*. Abbotsford: John Holland, http://www.johnholland.com.au/Documents.asp?ID=13993&Title=Safety.

Johnson, S. (2007). The predictive validity of safety climate. *Journal of Safety Research, 38*(5), 511–521. doi: 10.1016/j.jsr.2007.07.001

Keil Centre. (2011). Safety Culture Maturity® Model Retrieved 4 August, 2011, from http://www.keilcentre.co.uk/safety-culture-maturity-model.aspx

Lardner, R. (2002). Towards a mature safety culture. Paper presented at the The Institution of Chemical Engineers Conference. http://www.keilcentre.co.uk/human-factors-in-safety.aspx

Lee, T., & Harrison, K. (2000). Assessing safety culture in nuclear power stations. *Safety Science, 34*(1–3), 61–97.

Loughborough University. (2009). Safety Climate Measurement User Guide and Toolkit Retrieved from http://www.lut.ac.uk/departments/bs/safety/index.html

Mohamed, S. (2002). Safety climate in construction site environments. *Journal of Construction Engineering and Management, 128*(5), 375–384.

Neal, A., Griffin, M. A., & Hart, P. M. (2000). The impact of organizational climate on safety climate and individual behavior. *Safety Science, 34*(1–3), 99–109.

Ostrom, L., Wilhelmsen, C., & Kaplan, B. (1993). Assessing safety culture. *Nuclear Safety, 34*(2), 163–172.

Parker, D., Lawrie, M., & Hudson, P. (2006). A framework for understanding the development of organisational safety culture. *Safety Science, 44*(6), 551–562. doi: 10.1016/j.ssci.2005.10.004

Pepe, J., & Cataldo, P. J. (2011). Manage risk, build a just culture. *Health Progress*, (July–August), *92*(4), 56–60.

Pungvongsanuraks, P., & Chinda, T. (2010). Investigation of safety perceptions of management and workers in Thai construction industry. *Suranaree Journal of Science and Technology*, *17*(2), 177–191.

Reason, J. (1997). *Managing the Risks of Organisational Accidents.* Aldershot: Ashgate.

Reason, J. (1998). Achieving a safe culture: Theory and practice. *Work & Stress*, *12*(3), 293–306. doi: 10.1080/02678379808256868

Reason, J. (2000). Safety paradoxes and safety culture. *Injury Control and Safety Promotion*, *7*(1), 3–14.

Root, N. (1981). Injuries at work are fewer among older employees. *Monthly Labor Review*, *104*, 30–34.

Seo, D., Torabi, M., Blair, E., & Ellis, N. (2004). A cross-validation of safety climate scale using confirmatory factor analytic approach. *Journal of Safety Research*, *35*(4), 427–445. doi: 10.1016/j.jsr.2004.04.006

Sherratt, F. (2014). Exploring the utility of construction industry 'safety culture' Paper presented at the CIB W099 Achieving Sustainable Construction Health and Safety, Lund, Sweden.

Siu, O.-l., Phillips, D. R., & Leung, T.-w. (2004). Safety climate and safety performance among construction workers in Hong Kong: The role of psychological strains as mediators. *Accident Analysis and Prevention*, *36*(3), 359–366.

Sunindijo, R. Y., & Zou, P. X. W. (2012). Political skill for developing construction safety climate. *Journal of Construction Engineering and Management*, *138*(5), 605–612.

Sunindijo, R. Y., & Zou, P. X. W. (2013). Aligning safety policy development, learning and implementation: From boardroom to site Paper presented at the CIB World Building Congress 2013, Brisbane, Australia.

Varonen, U., & Mattila, M. (2000). The safety climate and its relationship to safety practices, safety of the work environment and occupational accidents in eight wood-processing companies. *Accident Analysis & Prevention*, *32*(6), 761–769.

Weick, K. E., Sutcliffe, K. M., & Obstfeld, D. (1999). Organizing for high reliability: Processes of collective mindfulness. In B. Staw & R. Sutton (Eds.), *Research in Organizational Behavior* (Vol. 21, pp. 23–81). Stamford, CT: JAI Press.

Wiegmann, D. A., Zhang, H., von Thaden, T., Sharma, G., & Mitchell, A. (2002). *A synthesis of safety culture and safety climate research.* Savoy: University of Illinois.

Williamson, A. M., Feyer, A., Cairns, D., & Biancotti, D. (1997). The development of a measure of safety climate: The role of safety perceptions and attitudes. *Safety Science*, *25*(1–3), 15–27.

Zhou, Q., Fang, D., & Mohamed, S. (2011). Safety climate improvement: Case study in a Chinese construction company. *Journal of Construction Engineering and Management*, *137*(1), 86–95. doi: 10.1061/(asce)co.1943-7862.0000241

Zohar, D. (1980). Safety climate in industrial organizations: Theoretical and applied implications. *Journal of Applied Psychology*, *65*(1), 96–102.

Zou, P. X. W. (2011). Fostering a strong construction safety culture. *Leadership and Management in Engineering*, *11*(1), 11–22.

Zou, P. X. W., & Sunindijo, R. Y. (2010). Construction safety culture: a revised framework. Paper presented at the The Chinese Research Institute of Construction Management (CRIOCM), 15th annual symposium, Johor Bahru, Malaysia.

4 Skills for Safety

This chapter focuses on key skills that project management personnel need to lead safety management in construction and engineering projects. Project management personnel are safety leaders on work sites. They play a vital role in developing and implementing safety strategies at the project level to ensure that the strategies are aligned with the safety mission and goals established at the higher management levels. In order to perform this role successfully and effectively, project management personnel require a range of skills. In this chapter we discuss the essential skills for project management personnel, which include conceptual, human, political and technical skills; we also propose a skill development model to assist project management personnel in developing such skills; we debate the applicability of the model in practice and recommend strategies to develop key skills.

An overview of the skill set

The commitment and participation of project management personnel as a key factor for the successful implementation of safety programmes has been articulated in numerous studies (Abudayyeh et al., 2006; Aksorn & Hadikusumo, 2008; Anton, 1989; Hudson, 2007; Zou & Sunindijo, 2010). Effective supervisory behaviour is proven to promote better safety (Mattila et al., 1994), while better performance can also be expected when site managers and supervisors engage workers in regular on-site safety talks (Langford et al., 2000). In essence, project management personnel have an important and ongoing safety leadership role and are responsible for performing safety management tasks to lead safety implementation in their projects (Dingsdag et al., 2006).

Strategic Safety Management in Construction and Engineering, First Edition.
Patrick X.W. Zou and Riza Yosia Sunindijo.
© 2015 John Wiley & Sons, Ltd. Published 2015 by John Wiley & Sons, Ltd.

Generally speaking, managers need to possess sufficient skills to perform their roles in organisations effectively. Management theoreticians have proposed a range of managerial skills over the past 50 years or so. Katz (1974) is one of the pioneers who investigated effective managerial skills. He examined the skills which executives exhibited to carry out their jobs effectively and suggested three basic developable skills, namely, (1) technical skill, (2) human skill, (3) conceptual skill. He defined technical skill as the 'specialised knowledge, analytical ability within that speciality and facility in the use of the tools and techniques of the specific discipline'; human skill as 'the ability to work effectively as a group member and to build cooperative effort within the team'; and conceptual skill as the 'ability to see the enterprise as a whole including recognising how the various functions depend on one another and how changes in one part can affect all the others' (pp. 91– 93). Katz added that conceptual skill extends to visualising the relationships of the business to the industry, the community, and the political, social, and economic conditions as a whole. This three-skill approach was ground-breaking at the time and is still prominent even today to the extent that it has been discussed and included in various management publications. Peterson and Fleet (2004) examined 15 management principles books published in the mid-1980s and 15 management textbooks published in the early 2000s. They found that 'Katz's work was specifically referenced by almost all of the early works and by most of the more recent books' (p. 1301). Management literature, both classic and contemporary, fundamentally agrees that there are three essential management skills: conceptual, human and technical.

In relation to safety management in construction and engineering, we contend that political skill is another set of skill that should be included. In today's workplace, it is essential to know what to do and how to do it in genuine, sincere and convincing ways. An individual needs to know when and how to put himself or herself in a proper place on certain issues to create and take advantage of opportunities (Ferris et al., 2005b). In other words, managers need to play politics in order to be successful at what they are doing. Several studies have indicated that the effective use of political skill is important for a person's career. This is because organisations are political arenas where competing interests, limited resources, coalition building and the exercise of power and influence thrive in getting things done. In fact, the exercise of politics is often one of the prime driving forces in organisations, for better or worse (Ferris et al., 2000; Pinto, 2000).

Traditionally, organisations have adopted a bureaucratic style with a tall hierarchical structure and a formalised chain of command. Globalisation, downsizing, restructuring, mergers, acquisitions, and the application of new technologies have changed the ways in which organisations function (Ferris et al., 2005b). The nature of organisations has become more social. Individuals do not spend their time working on individual tasks that separate them from others; rather, coordination and cooperation are required to achieve organisational goals. Unstructured interactions among team members, subordinates, peers and supervisors have become common occurrences. In current working environments, individuals are not considered to be good at their jobs if they are not experts at working with, and influencing, others (Brouer et al., 2009; Ferris et al., 2000; Ferris et al., 2005b; Smith et al., 2009). Consequently politics has become an integral part of every organisation where individuals interact with each other to resolve conflicts, share limited amounts of resources, and gain greater power (Vigoda, 2003). Political skill has become a necessity in current dynamic environment to advance personal and organisational agendas (Buchanan & Badham, 1999; Holden, 1998).

The conduct of politics is also inevitable in construction project management. Many successful project managers understand the importance of maintaining strong political ties as a method of attaining project success. They are aware that politics, used prudently, can have a positive impact on the implementation of their projects (Pinto, 2000). Blickle et al. (2009) found that politically skilled individuals are more likely to succeed in the enterprising job environment which includes more ambiguous environments, allowing individuals to facilitate multiple roles, and is characterised by the extensive use of verbal facility to persuade other people. This environment requires the ability to relate to a wide range of people across a variety of situations through talking and listening. This exactly describes what happens in the construction business environment. Construction is known as an ambiguous work environment because of the wide variety of components that have to be managed. Various external and uncontrolled factors and forces may create uncertainties and influence the achievement of business objectives. Furthermore, the involvement of internal and external stakeholders compels the management team to build relationships with people who have different agendas and backgrounds.

Based on the above, we propose four sets of essential skills, comprising *conceptual skill, human skill, political skil,* and *technical skill,* for project management personnel to perform their safety leadership roles. It is clear that Katz's (1974) work serves as the foundation in the development of the skill sets because his three-skill approach has been widely accepted. Building on this, we add political skill as the fourth essential skill to accommodate the need for such skill in the contemporary workplace. We then reviewed 16 studies on managerial skills from the fields of general management, project management, and construction management. We found that the skill components proposed in these 16 studies can be organised into the proposed four-skill construct. Table 4.1 summarises these studies and the skill components that they identified.

Based on these previous studies and literature, Sunindijo and Zou (2011) summarised the components of each skill as presented in Figure 4.1. These four sets of skills, including their components and their relevance to safety, are discussed in the following sections.

Table 4.1 Studies on managerial skills

No	Authors	Conceptual skill	Human skill	Political skill	Technical skill
A	*General management*				
1	Katz (1974)	Conceptual	Human	N/A	Technical
2	Peterson and Fleet (2004)	– Analytic – Decision making – Conceptual – Diagnostic – Ability to deal with ambiguous and complex situations	– Human – Communication – Interpersonal	N/A	– Technical – Administrative
B	*Project management*				
3	Gushgari et al. (1997)	– Decision making – Problem solving	– Communication – Listening – Leadership – Motivation	N/A	– Project management knowledge
4	El-Sabaa (2001)	– Planning – Organising – Strong goal orientation – Ability to see the project as a whole – Ability to visualise the relationship of the project to the industry and community – Problem solving	– Mobilising – Communication – Coping with situations (implying adaptability) – Delegating authority – Self-esteem Enthusiasm	– Political sensitivity	– Knowledge in using tools and techniques – Project knowledge – Understanding methods, procedures, and processes – Specialised technology – Computer skills
5	Lientz and Rea (2002)	– Generalist – Problem solving – Big picture perspective – Organising – Understanding how the organisation functions	– Communication – Conflict resolution – People management – Listening	N/A	– Technical knowledge – Specialised technology – Managing electronic communication

(continued overleaf)

Table 4.1 (*continued*)

No	Authors	Conceptual skill	Human skill	Political skill	Technical skill
6	Lei and Skitmore (2004)	– Legal understanding – Business knowledge	– People skill – Client-related skill – Stakeholder management skill – Coaching skill	– Networking skill	– Technical skill – Cost management – Computing skill – Risk management – Time management
7	Gillard and Price (2005)	– Diagnostic use of concepts – Efficiency orientation	– Self-confidence – Oral communication – Managing group process – Developing others – Adaptability	– Use of socialised power – Use of unilateral power (different authorities and power to influence others)	N/A
8	Brill et al. (2006)	– Problem solving – Analytical expertise	– Leadership – People management – Communication	N/A	– Context knowledge – Project administration – Use of tools
9	Kerzner (2009)	– Planning – Organising – Entrepreneurship – Resource allocation	– Team building – Leadership – Conflict resolution – Stakeholder management	N/A	– Technical – Administrative
10	Project Management Institute (2013)	– Decision making	– Leadership – Team building – Motivation – Communication	– Influencing – Negotiation – Political and cultural awareness	– Project management knowledge
C	*Construction management*				
11	Goodwin (1993)	– Conceptual	– Human	– Negotiation	– Technical
12	Odusami (2002)	– Decision making – Problem solving	– Communication – Leadership – Motivation	N/A	N/A

13	Dainty et al. (2003)	– Decision making – Assimilating information and using it to formulate actions	– Team building – Leadership – Mutuality and approachability – Communication – Self-efficacy	– Being perceived as honest and having integrity – Dealing with external relations outside the project team	N/A
14	Cheng et al. (2005)	– Achievement orientation – Proactive – Analytical thinking – Conceptual thinking	– Focus on client's needs – Leadership – Self-control – Flexibility	– Impact and influence – Directiveness to ensure subordinate compliance – Influencing team performance	N/A
15	Chen et al. (2008)	– Planning – Commercial management – Ability to coordinate	– Communication – Team management – Relationship building	N/A	– Knowledge of construction work
16	Farooqui et al. (2008)	– Creativity – Decision making – Organising – Planning and goal setting – Problem solving – Ability to follow up – Critical path thinking – Result orientation	– Communication – Delegating – Leadership – Motivation – Listening	– Negotiation – Personal adaptability	– Administrative – Financial management – Quality management – Risk management – Technical knowledge – Time management

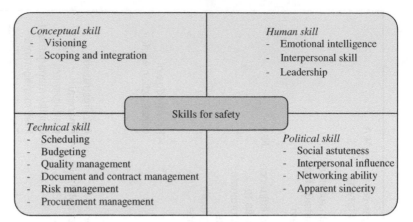

Figure 4.1   Skills for safety in construction and engineering

Conceptual skill

Generally speaking, conceptual skill is referred to as the ability to see an enterprise as a whole (Katz, 1974). It is lauded as the unifying, coordinating ingredient of the administrative process, and has undeniable overall importance. In a construction business and project for example, conceptual skill is crucial for viewing the project from a big picture perspective, to understand the dynamic relationships among different project components and stakeholders, and to envision how the project affects its surrounding environment (El-Sabaa, 2001; Goodwin, 1993). Due to the diversity of construction project systems, conceptual skill is paramount in ensuring that the systems function as an integrated whole (Goodwin, 1993). When there is a change in one system, the impacts should be considered against different indicators, such as control strategies, budget, schedule, and environment (Katz, 1974). This conceptual skill is also useful for viewing the project as one of many inter-related projects within the organisation.

In the context of safety management, project management personnel need conceptual skill to appreciate the impacts and necessities of good safety practices towards the workers and their families, the organisation, the community, and the achievement of project objectives. Conceptual skill provides an overall view and understanding on the roles of safety in a construction project, thus helping project management personnel realise that safety is actually an integral part of the project. When project management personnel perceive the importance of safety, they are motivated to act in ways that advance safety.

The ability to see the interrelationship between different components within a project through the application of conceptual skill gives project management personnel a strategic understanding of implementing safety measures required in the project. It ensures that proper safety measures are implemented so that these are neither too excessive or impractical that they increase the cost unreasonably and delay the project, nor too relaxed that unsafe acts and conditions are flourishing. Furthermore, conceptual skill not only enables project management personnel to identify various project components, but also helps them identify safety hazards and risks throughout the construction life cycle. This allows them to develop and incorporate a safety plan into the overall construction plan (Sunindijo & Zou, 2013).

Components of conceptual skill

Based on Table 4.1, researchers have adopted different names to explain the components of conceptual skill. Based on the definitions and rationalisations given in previous studies, they can be summarised into seven components:

- *Visualising.* The ability to identify key aspects in the project and their interrelationship.
- *Decision making and prioritising.* The ability to prioritise and make decisions based on available alternatives and the achievement of overall organisational/project objectives.
- *Problem diagnosing.* The ability to identify the root cause of problems.
- *Systemic problem solving.* Solving problems from a system-wide perspective, considering the impacts of the solutions on the organisation/project as a whole.
- *Planning.* The ability to define objectives and decide on the tasks and resources needed to attain them.
- *Organising.* The ability to distribute resources and decide on the roles and responsibilities of personnel.
- *Goal orientation.* The ability to constantly fix on set goals in planning, organising, making decisions, performing tasks and solving problems.

Using questionnaire surveys, we collected data from large construction and engineering organisations in Australia. Our analysis results showed that the components of conceptual skill can be further summarised into two components, namely, visioning, and scoping and integration (Sunindijo & Zou, 2013). Visioning is the ability to view and understand the project/organisation as a whole and to use that understanding to make decisions that promote the achievement of key objectives. Scoping and integration is the ability to determine and control what needs to be included in a scope of work and to ensure that all components are properly identified, combined, unified and coordinated.

Examples of the manifestation of visioning in the construction and engineering context are:

- Organising people, which includes defining roles and responsibilities, determining who reports to whom and at what level decisions should be made.

- Prioritising or making trade-offs among competing objectives and alternatives.
- Evaluate performance against standards and goals.
- Taking corrective actions to improve performance as necessary.

In the same way, examples of the manifestation of scoping and integration in the construction and engineering context are:

- Preparing a master schedule for a construction project.
- Estimating an overall project cost including project profit plan.
- Understanding of relationships among work packages.
- Understanding of contractual agreements and their risks imposed on the project.
- Making decisions from a system-wide perspective and understanding the impacts of the decisions on various project elements.

Conceptual skill and safety management

We also developed a model to visualise the relationship between conceptual skill and safety management as shown in Figure 4.2. The visioning component of conceptual skill is an initiator of the whole relationship. In a construction project, project team members must coordinate and work together to achieve project objectives. There must be a common vision to serve as guidelines while clear expectations of the work need to be determined (Ellis, 2005). Visioning is also about prioritising among competing objectives and alternatives. There are various goals, targets, and expectations, which may distract project management personnel from the key objectives of the project. Therefore, they must be able to discern major priorities and remain focused on these priorities in decision making. They need to be aware of how certain decisions can affect other project objectives and recognise problems that may arise during the implementation of these decisions. This understanding is a key to identify and evaluate a set of viable alternatives (Chung & Megginson, 1981), thus enabling project management personnel to select the best alternative for the project.

The visioning component is a predictor of the scoping and integration component. When a clear vision and a set of objectives have been determined, project management personnel have a foundation to define their scope of work more accurately. They will be able to develop a realistic schedule and prepare a

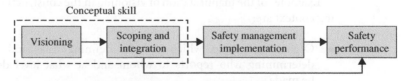

Figure 4.2   Conceptual skill and safety management

budget to achieve project objectives as stated in contractual agreements. Having clear objectives also helps project management personnel identify various work packages in the project and how they relate to one another. For example, the foundation work of a building must be completed before the upper-structure work can be commenced. Therefore, if the foundation work is delayed, the upper-structure work will be delayed as well. It is likely that the foundation work is one of the critical activities, thus if it is delayed, it will delay other works, such as the external work, mechanical and electrical work, and interior work, which will eventually delay the completion date of the whole project. When project management personnel have a clear understanding of these relationships, they can make informed decisions from a system-wide perspective. They would consider the relationships between work packages, the requirements stated in the contracts, the expectations of project stakeholders, and other aspects that may influence project success, before they make those decisions.

The relationship between scoping and integration and safety management implementation indicates that safety should be considered by project management personnel in their planning. For example, when preparing a project schedule, safety management must become an integral part of the procedures. The schedule must take into account the necessary time to conduct regular safety training, safety inductions and safety meetings. Safety risks can also be attached to each activity in the project schedule, thus the right resources can be allocated, safety constraints can be considered and alleviated in advance and safety control can be improved (Wang et al., 2006). Likewise, the project budget must include necessary proportions on safety investments, such as personal protective equipment, safety training and other safety measures, to support the implementation of safety management. The implementation of safety management will then promote better safety performance.

Figure 4.2 also shows that scoping and integration directly influences safety performance. This is understandable because when project management personnel include safety in the project schedule and budget, as well as incorporating safety requirements in contract agreements or as part of tender requirements, the other stakeholders will realise that safety is being regarded as important in the project, thus positive attitudes and perceptions towards safety will be developed. Furthermore, when safety becomes part of project management personnel's responsibilities and when safety is included as part of the responsibilities of others, the whole project team will be more committed to safety, leading to better safety performance.

Human skill

Human skill is referred to as the ability to work with and through other people (Goodwin, 1993; Katz, 1974), which is understandably crucial in a construction business due to the involvement of various stakeholders and its labour-intensive work nature. This is also true when it comes to safety where project management personnel depend on the others to perform the work in a safe manner.

Components of human skill

Based on previous literature as summarised in Table 4.1, three components of human skill have been identified as important, particularly in their contribution to safety management. The three components are emotional intelligence, interpersonal skill, and leadership.

Emotional intelligence is 'the capacity for recognising our own feelings and those of others, for motivating ourselves, and for managing emotions well in ourselves and in our relationships' (Goleman, 1998, p. 317). It is closely associated with superior performance in various industries, including the construction industry. There are four dimensions of emotional intelligence. The first dimension is self-awareness or the ability to accurately perceive one's own emotions and remain aware of them as they happen. This ability makes people aware of how they tend to respond to specific situations and people. The second dimension is self-management, which is the ability to use awareness of emotions to stay flexible and direct own behaviour positively. This dimension is about managing emotional reactions to all situations and people. The third dimension is social awareness, which is the ability to accurately recognise emotions in others and understand what they think and feel. The last dimension is relationship management, which is the ability to use awareness of one's own emotions and the emotions of others to manage interactions successfully (Bradberry & Greaves, 2001–2010). Relationship management is the culmination of the other dimensions of emotional intelligence. It is needed by project management personnel to nurture positive relationships with various stakeholders during the implementation of safety management strategies.

Interpersonal skill, the second component of human skill, refers to the ease and comfort of communication between individuals and their colleagues, superiors, subordinates, clients, and other stakeholders (Peled, 2000). This skill is essential to communicate to others about safety and to motivate people during safety implementation. Furthermore, interpersonal skill is a key to promoting teamwork and resolving conflicts that may hinder the attainment of safety goals. Interpersonal skill is also a dominant factor that influences leadership effectiveness (the third component of human skill which is discussed in the next point). As safety leaders, project management personnel have to exercise their interpersonal skill in their day-to-day interactions with others, particularly to provide safety leadership. In short, there is no possibility of leadership without these interactions being effectively managed.

Leadership, as the third component of human skill, refers to the ability to obtain followers (Drucker, 1996; Maxwell, 1993). Leadership theories have evolved from the trait theories to the contingency theories. Today, transformational leadership is considered to be particularly effective because it is able to stimulate and inspire followers to go beyond their own self-interest to achieve extraordinary outcomes for the good of the organisation (Robbins et al., 2012). Transformational leaders demonstrate four characteristics. First, they provide idealised influence by behaving in ways that allow them to serve as role models for their followers, thus making them admired, respected, and trusted (transformational leaders are endowed by their followers with extraordinary capabilities, persistence and determination). Second, they provide inspirational motivation, meaning and challenge to their followers' work (transformational leaders clearly communicate expectations and demonstrate commitment to the shared vision and goals). Third, they provide intellectual stimulation and encourage followers to be innovative and creative by questioning assumptions, reframing problems and approaching old situations in new ways. Fourth, they provide individualised consideration by giving special attention to each individual follower's needs for achievement and growth by acting as a coach or mentor (Bass & Riggio, 2006). Transformational leadership is an important management tool to provide safety leadership during the implementation of safety strategies both in organisational and project levels. Using transformational leadership as a basis, project management personnel should become role models to build safety commitment. They can inspire others by articulating a clear vision and showing the moral values of safety, thus increasing the intrinsic value of achieving safety goals. Such a charismatic approach should be supported by necessary training and mentoring to provide others with a sense of increased competence to carry out safety management responsibilities. This creates more satisfied followers, while simultaneously promoteing positive perceptions and attitudes towards safety.

Human skill and safety management

Based on the analysis of data we collected from the construction industry in Australia, we developed a model to visualise the relationship between human skill and safety management as shown in Figure 4.3. Emotional intelligence is the foundation of the relationships shown in the model. Within the emotional intelligence component, self-awareness is the prerequisite of the other three dimensions, that is, self-management, social awareness, and relationship management. Self-awareness is an ability that has been appreciated since ancient times. Individuals high in self-awareness understand their strengths and limitations, seek feedback, learn from their mistakes, and understand when

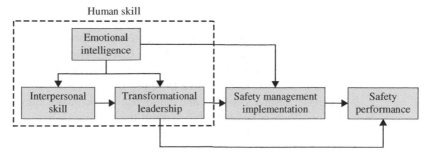

Figure 4.3   Human skill and safety management

to work with others who have complementary strengths. This understanding of oneself brings about effective self-management and greater understanding of others, making these individuals appear trustworthy and competent (Janasz et al., 2006). Next, self-management is a predictor of social awareness and relationship management. The relationships are evident, given that individuals who cannot control their emotional outbursts will have less chance to be effective in understanding others and developing relationships (Goleman, 2001). Research involving children with autism found that self-management can be used to improve social skills (Koegel et al., 1992), while at a neurological level, self-management is a foundation of social effectiveness (Damasio, 1994). Lastly, social awareness is a predictor of relationship management. Lane (2000) suggested that understanding of one's own emotions and the emotions of others is a way to create effective social interactions. Individuals with high social awareness understand different points of view, making them effective in their interactions with different types of people. As a result, they are able to get along in organisational life, build networks, and employ influence tactics to achieve positive results (Robbins & Hunsaker, 2009).

Emotional intelligence is a prerequisite of effective application of interpersonal skill. The self-management dimension enables individuals to manage emotional outbursts, which is essential for communicating effectively, resolving conflicts, and building teamwork (Janasz et al., 2006). The social awareness dimension is critical for superior job performance whenever the focus is on interactions with people. Furthermore, socially aware individuals are emphatic, having an ability to put themselves in someone else's shoes, sense their emotions, and understand their perspective, thus enabling them to interact effectively with different types of personalities (Robbins & Hunsaker, 2009). The relationship management dimension is useful for developing, coaching, mentoring and persuading others to achieve common goals. It generates effective communication and conflict management, which enhance networking, collaboration and teamwork (Goleman, 2001; Robbins & Hunsaker, 2009).

Emotional intelligence and interpersonal skill are predictors of transformational leadership. Goleman (2001) suggested that individuals competent in relationship management, a dimension of emotional intelligence, are able to sense the developmental needs of others, making them excellent coaches and mentors. They are influential and articulate a shared vision that arouses enthusiasm and inspires others to work together towards common goals. They are also change catalysts who bring greater efforts and better performance from their subordinates. This shows that emotional intelligence generates competencies required by individuals to be transformational leaders. Likewise, transformational leaders have to apply interpersonal skill to communicate, motivate, resolve conflicts and build teamwork before they can be effective leaders. There is no leadership without these interactions.

When it comes to safety management implementation, the self-management and relationship management dimensions of emotional intelligence are essential. Self-management can be considered as a form of self-leadership where individuals motivate themselves to achieve their goals (Robbins & Hunsaker, 2009). In practising self-management, project management personnel should include safety as one of their values and goals. This, in effect, will influence their decisions and behaviour, motivating them to focus on safety amid setbacks and difficulties. Relationship management, on the other hand, is needed to relate with others in ways that nurture positive relationships, thus enabling project management personnel to make the most out of every interaction (Bradberry & Greaves, 2009). This is important because it is impossible for project management personnel to implement safety management on their own. They need to collaborate with and be supported by other project stakeholders.

Finally, it is interesting to note that there is a direct relationship between transformational leadership and safety performance. Transformational leadership helps project management personnel to be role models to build safety commitment. They need to inspire others by articulating a clear vision and showing the moral values of safety, thus increasing the intrinsic value of achieving safety goals. Such a charismatic approach should be supported by necessary training and mentoring to provide others with a sense of increased competence to carry out safety duties. This creates more satisfied followers, while simultaneously promoting positive perceptions and attitudes towards safety, which improves safety performance (Sunindijo & Zou, 2014).

Political skill

Political skill is referred to as 'the ability to understand others at work and to use that knowledge to influence others to act in ways that enhance one's personal and organisational objectives' (Ferris et al., 2005b, p. 7). As stated earlier, political skill has become a necessity in the current dynamic environment to advance personal and organisational agendas because politics has become an integral part of every organisation where individuals interact with each other to resolve conflicts, share limited amounts of resources and gain greater power (Vigoda, 2003). Politically-skilled managers are those who expect to experience resistance to their attempts to get things done, but nevertheless keep on taking initiatives, carefully selected initiatives, in ways that eventually tend to produce the results they desire (Hayes, 1984).

Many people are unclear about the differences between political skill and human skill. As an illustration, due to their human skill, some managers are loved and admired by many. However, they fail when, for example, managing a project. A reason behind this failure is a lack of political skill, which is necessary to manipulate interpersonal relationships with others to ensure the ultimate success of the project (Peled, 2000). The main difference between political and human skills is that political skill is specific to interactions aimed to achieve success in organisations. None of the previous forms of human skill was developed explicitly to address interpersonal interactions in organisational settings (Ferris et al., 2000). Furthermore, human skill generally refers to competencies in communication and the ability to interact with others. Political skill, on the other hand, goes beyond mere ease and facility of interaction. It focuses on managing interactions with others in influential ways that lead to individual and organisational goal accomplishments (Ferris et al., 2005b). Research has also found evidence that political skill is empirically distinct from human skill components (Ferris et al., 2005c).

Research has confirmed the value of political skill in organisations. For example, Mainiero (1994) found that political skill is an important factor that contributes to career advancement of women, while Perrewé and Nelson (2004) argued that political skill is essential for successful female managers. Spencer and Spencer (1993) suggested that well-developed political skill is an important contributor that distinguishes superior performers. Ahearn et al. (2004) found that a leader's political skill causes teams to perform more effectively. Ferris et al. (2000) added that people with high political skill know what to do in different social situations at work and how to do it in a sincere manner, thus disguising any potential manipulative motives and rendering the influence attempts successful. They convey a sense of personal security and calm self-confidence that attracts and gives others a feeling of comfort. This self-confidence is displayed at the proper level, thus it is seen as a positive attribute instead of coming across as arrogance. They are not self-absorbed, but their focus is outwards towards others, instead of inwardly and self-centred. Politically skilled individuals may have ulterior and self-serving motives, but their behaviour will always be the same, regardless of their underlying motives. Without the skill, people can be completely sincere and devoted to the common good and still find that others doubt their motives and therefore withdraw from them (Ferris et al., 2005b). Recent cross-sectional and longitudinal research has shown that political skill accounted for a significant proportion of job performance variance beyond general mental ability and personality (Blickle et al., 2011).

Components of political skill

There are four components of political skill: social astuteness, interpersonal influence, networking ability and apparent sincerity. Social astuteness is the first component. Individuals with high political skill are astute observers of others and keenly attuned to diverse social situations. They are also sensitive to others, and are, thus, considered as ingenious and clever in dealing with others.

The second component is interpersonal influence, which is the convincing personal style that exerts a strong influence on people around them. Individuals with good political skill are flexible and can appropriately adapt their behaviour to each situation in order to extract certain responses from others. This component is different from interpersonal skill of the human skill construct, although there are some overlapping aspects between the two. Interpersonal skill is the ability to build relationships and get along with others, whilst interpersonal influence is a tool to manipulate interpersonal relationships to ensure the ultimate success of a project or organisation (Peled, 2000). Individuals high in interpersonal influence not only appear to others as being pleasant and productive, but they have the ability to control their environments (Ferris et al., 2005b).

The third component of political skill is networking, which means the ability to develop and use diverse networks of people or networking ability. People included in the networks are considered to hold assets deemed as valuable and necessary for attaining successful personal and organisational functioning. People with high networking ability are often expert negotiators, deal makers and at ease with conflict management (Ferris et al., 2005b).

The fourth political skill component is apparent sincerity. This competency is the key to influence others because it focuses on the perceived intentions of certain behaviour exhibitions. The influence attempts will be successful when there are no ulterior motives behind the behaviour exhibited. People high in apparent sincerity inspire trust and confidence because they do not appear to be manipulative or coercive (Ferris et al., 2005a; Ferris et al., 2005b; Ferris et al., 2007) whether or not they have hidden agendas. On the other hand, people lacking in political skill may be sincere, but others still doubt their motive.

Political skill and safety management

Project management personnel will need to demonstrate genuine interests towards safety by exercising their political skill. This will influence project team members and other stakeholders in recognising the importance of safety in the project and convince them to consider safety equally important as the other project objectives (Sunindijo & Zou, 2012a). Furthermore, project stakeholders may, at times, be unwilling to offer their help and support on safety unless they perceive that it is in their interests to do so. As such, project management personnel need to use political skill to cultivate relationships with the power holders and make the deals that are needed to improve safety (Pinto, 1998; Sunindijo & Zou, 2012a). Political skill also enables project management personnel to adapt their behaviour and influence tactics to suit others, thus inducing others to implement safety strategies for generating and maintaining a safe work environment.

We developed a model to visualise the relationship between political skill and safety management as shown in Figure 4.4. Among the components of political skill, apparent sincerity can be considered as the foundation. First impressions are crucial in any relationships because they influence the way people see subsequent data about the perceived object or person. Making favourable first impressions is important in every socialisation process (Chung & Megginson, 1981). In the context of safety management, it is important for project management personnel to appear sincere and show genuine interests towards safety as part of any first impressions. As a consequence of this, other project stakeholders will recognise the importance of safety and project management personnel will have a positive start in influencing others to consider safety to be as important as other project objectives. Essentially,

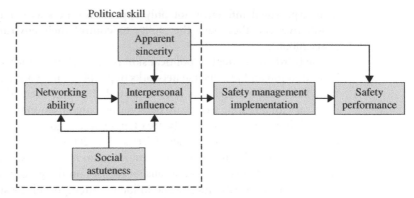

Figure 4.4   Political skill and safety management

apparent sincerity is the core that makes political skill work because it allows individuals to exert influence in a way that does not create ill-will or undue influence, thus others do not question their motives or react negatively (Ferris et al., 2005a).

The model further shows that social astuteness is a predictor of networking ability. Socially astute individuals are experts in social interactions and accurately interpret their own behaviour and those of others in social settings (Ferris et al., 2005b). This understanding of others and how interpersonal interactions work makes socially astute project management personnel able to determine the right approach to build and maintain networks, with influential people who can help them achieve personal and project goals.

Another set of relationships shows that social astuteness, networking ability and apparent sincerity are the prerequisites of interpersonal influence. Individuals with high interpersonal influence are flexible and can appropriately adapt their behaviour to various situations to elicit the desired responses from others, thus allowing organisational or project goals to be attained (Ferris et al., 2000; Ferris et al., 2005b). In order to influence others in the construction and engineering context, project management personnel need to have a good understanding of others and know how a project is managed, by using their social astuteness. This helps project management personnel analyse and employ the right influence tactics beneficial for the projects (Buchanan & Badham, 1999). Apparent sincerity is also needed because, as stated earlier, this component is the core of political skill and project management personnel have to appear genuine without any hidden agendas if they want to influence others to do something. By developing networks through networking ability, it will be much easier to influence people who are already in the networks or to use those networks to influence others. Individuals with high networking ability are often expert negotiators, deal makers and at ease with conflict management (Ferris et al., 2005b).

This interrelationship among the components of political skill aligns with an argument put forward by Holden (1998) who stated that individuals who practice influence have the following qualities: the ability to see from a big picture perspective (understand organisational objectives and strategies); having a clear insight of the role of different people to achieve objectives; the ability to align personal goals with others and a non-threatening appearance. In this case, social astuteness is the key to seeing the whole picture of a construction project, understanding the roles of various stakeholders, and identifying social interactions and alliances within the project. Using networking ability, project management personnel can develop relationships with others, thus finding common ground or alignment to work together and achieve project objectives. Lastly, non-threatening appearance is related to apparent sincerity, where project management personnel seem honest and forthright, without appearing manipulative and coercive.

Interpersonal influence is the final step in which project management personnel should apply all the knowledge and understanding (concerning others and the way the project is run) that they have accumulated, to determine the right strategy to influence others. This component of political skill helps project management personnel increase their effectiveness in leading the implementation of safety management, which promotes better safety performance. Project management personnel should use their interpersonal influence when implementing safety management to make others willing to follow safety procedures and perform safely. Interpersonal influence enables them to adapt their behaviour and influence tactics to suit others; thus they can influence others to implement the safety management tasks to both create and maintain a safe working environment.

Apparent sincerity has a direct relationship with safety performance. When project management personnel appear sincere and always show genuine interest on safety implementation and issues, this behaviour and attitude will also affect the safety perceptions of others. People will feel that project management personnel are committed to safety, which will also make them have positive perceptions and attitudes toward safety. Many studies have found that this kind of management commitment is an important factor in improving safety performance (Sunindijo & Zou, 2012b).

Technical skills

Technical skill is referred to as the job-specific knowledge and techniques that are required to perform specific tasks proficiently (Robbins et al., 2012). It involves specialised knowledge, analytical ability within that speciality and facility in the use of the tools and techniques of the specific discipline (Katz, 1974).

Components of technical skill

Based on Table 4.1, six components of technical skill have been identified as particularly relevant in the construction and engineering context (Sunindijo & Zou, 2011).

The first component is scheduling, which involves an understanding in determining the dates when different activities will be performed, recognising activities that drive other activities and determining when the activities are due (Farooqui et al., 2008; Project Management Institute, 2013).

The second component is budgeting and cost management, which involves determining the types and quantities of resources needed to perform various project activities, developing cost estimation for all resources, allocating the budget to individual work activities and controlling changes to the project budget (Project Management Institute, 2013).

The third component is quality management, which includes activities such as identifying relevant quality standards and determining how to meet them, evaluating project performance periodically to provide confidence that the project will meet the standards and monitoring specific results to determine their compliance with the standards, as well as finding ways to eliminate unsatisfactory performance (Farooqui et al., 2008; Project Management Institute, 2013).

The fourth component is document and contract administration, an understanding of procedures for implementing construction contracts according to the accepted practices and regulations within the construction industry. In addition, it includes the setting up of a management system for keeping records and reports of daily activities (Fisk, 1997).

The fifth component is risk management, normally involving five steps: establishing the context, risk identification, risk analysis, risk evaluation and risk treatment (Australian Standards & New Zealand Standards, 2009). Every construction project involves risks, be it cost related, time related, quality related or safety and environmental related. It is important that project management personnel are equipped with risk management skills to manage project risks in order to achieve project objectives.

The sixth component is procurement management, which includes the processes required to obtain goods and services from outside the organisation or from external parties such as consultants, subcontractors, vendors, and suppliers (Project Management Institute, 2013).

In relation to safety management, project management personnel have to exercise their technical skill to ensure that all site activities are performed in a safe manner. For example, risk management skill is needed to identify, evaluate and manage safety risks. The risk management skill, budgeting skill and scheduling skill make project management personnel realise the severe impacts of an accident on their project. Using the procurement skill, project management personnel are able to evaluate tender submissions and award work packages to contractors who have offered a reasonable price along with good safety records. The document and contract administration skill is important to make sure that all safety-related documents, permits, audits, procedures, safe work method statements and policies are processed and distributed in a timely manner to all concerned people (Sunindijo & Zou, 2012a).

Technical skill and safety management

Figure 4.5 depicts the relationships between technical skill and safety management. Based on the data that we collected in the Australian construction industry, among the six components of technical skill, only three components have a direct influence on safety, namely, budgeting, document and contract administration, and risk management.

Document and contract administration skill refers to the ability to use an effective documentation system and procedure for performing daily activities and tracking various changes that may happen in the project. It also involves a proficiency in managing project documents, such as drawings, submittals, requests for information, safe work method statements, change orders and payment requests. This skill may seem trivial, but poor performance in this area could be devastating. For example, inaccurate payment requests or late responses to requests for information will disrupt the entire construction process. The work flow will suffer, the schedule and budget will be compromised and the entire project will experience unnecessary risks. The case is the same when it is applied to the implementation of safety management tasks. Document and contract administration is required to implement the majority of safety management related tasks. As an example, the task of developing safety procedures and instructions requires a capacity to create and develop process documents that comply with legislation and meet the organisation's values. Furthermore, these procedures and instructions must be available in a written format and distributed to relevant stakeholders for maintaining safety standards throughout the project (Dingsdag et al., 2006). Efficient procedures and proper documents are also required to prepare, submit, and approve safe work method statements. Poor document and contract administration, in this case, may lead to delayed responses or lost submissions which will delay the

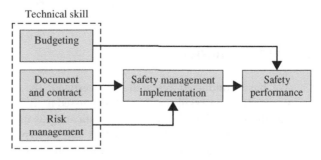

Figure 4.5   Technical skill and safety management

work activities and potentially affect the entire project schedule. Langford et al. (2000) found that providing every worker with a safety booklet or manual is important for developing positive attitudes towards safety. This implies that the document and contract administration skill is needed to prepare such a written safety programme and to distribute the programme to everyone involved.

Risk management is the core of safety management implementation. Dingsdag et al. (2006) argued that implementing safety management involves two task categories. The first category is to proactively identify, assess and determine appropriate controls for safety risks, while the second is to effectively communicate and consult with stakeholders regarding safety risks. The importance of the risk management skill is apparent in these two categories. The first category is about implementing the whole process of risk management. It is a proactive approach where safety risks are identified and managed in advance, before the construction stage begins. In particular, risk identification is a crucial step because unidentified risks negate the entire risk management process. If those involved are not aware of the risks in the first place, then the risks cannot be evaluated and, as a result, mitigation strategies cannot be planned and implemented (Carter & Smith, 2006). The second category is more process-oriented, where the tasks are mostly performed during the construction stage. This category is the continuation of the risk management process in which the risk management plan developed earlier is being put into practice. It also involves communicating safety risks to relevant stakeholders to make sure that each activity in the project is performed according to the safety plan and procedures. Risk management, therefore, should be considered as a continuous process throughout the construction life cycle (Akintoye & MacLeod, 1997).

Budgeting has a direct influence on safety performance. This is understandable as discussed in detail in Chapter 2. Higher amount of safety investment has yielded a return on investment of 46.08%, while, on the contrary, lack of safety has an adverse impact on the economic performance of a construction project because an accident can cost up to AU\$1.6 million. Without proper understanding of cost−benefit analysis, project management personnel may not appreciate the potential saving generated from implementing necessary safety measures. When proper budget for safety is allocated into each work package, it will enhance the stakeholders' perception concerning the organisation's safety commitment and, as a result, safety performance will also be improved.

Skill development model

We have conducted research to investigate the relationships between the components and skill components discussed in the previous section. Structural equation modelling (SEM) was the data analysis method used in the research. SEM is a statistical methodology that takes a hypothesis-testing approach to the analysis of a structural theory bearing on some phenomenon. This method allows a simultaneous examination of relationships among independent and dependent variables or constructs within a theoretical model. The numbers on the arrows are the path coefficients, which represent the amount of change in the dependent variables per single unit change in the predictor variables (Byrne, 2010). For example, the coefficient between visioning, and scoping and integration is 0.45, which suggests that for every single unit of increase in visioning, scoping and integration is increased by 0.45 unit, based on the data that have been collected. Figure 4.6 shows the actual model derived from the analysis. Figure 4.7 simplifies this model and highlights the most important skill components that influence safety management implementation in construction projects (Zou & Sunindijo, 2013).

Visioning (a component of conceptual skill), self-awareness (a component of emotional intelligence, as a component of human skill) and apparent sincerity (a component of political skill) are foundation and initiators of the entire relationship. Due to the involvement of project team members who may have different backgrounds and expectations, visioning is crucial in creating a common vision, which serves as a guide and becomes a common ground for every member to work collaboratively (Ellis, 2005). Safety should be part of the overall vision and one of the key project objectives. This visioning or foresight is an indispensable quality of leadership. A vision is the energy that pushes through all difficulties. It unites and induces people to sacrifice for accomplishing the

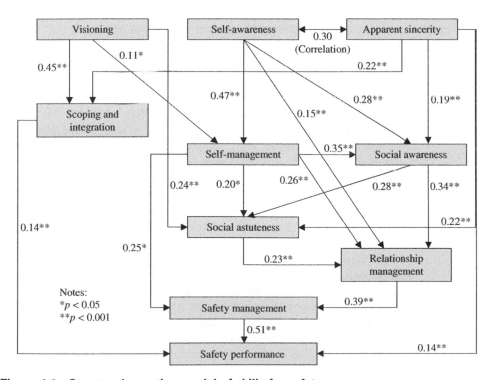

Figure 4.6   Structural equation model of skills for safety

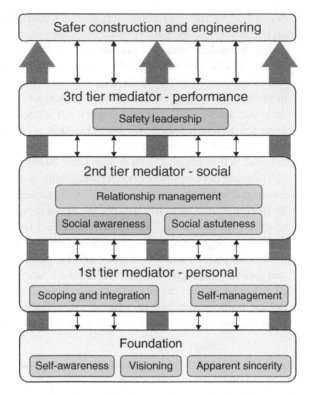

Figure 4.7   Skill Development Model (Zou & Sunindijo, 2013)

goal. A maxim which says 'What you see is what you get' cannot be any truer in relation to this concept of visioning (Maxwell, 1993).

The model illustrates that visioning is a predictor of scoping and integration (a component of conceptual skill), self-management (a component of emotional intelligence) and social astuteness (a component of political skill). The relationship between visioning, and scoping and integration is clear. With a clear vision and specific objectives, project management personnel will be able to determine the scope of the project, which will help them develop a realistic schedule and estimate the overall project cost (Project Management Institute, 2013). Visioning also helps project management personnel integrate various work packages and determine how they relate to one another. Furthermore, when project management personnel understand such relationships in the project, they can make decisions systematically and consider the relationships between work packages, the requirements stated in the contracts, the expectations of project stakeholders and other aspects that may influence project success, before they make any decision.

Another relationship evident in the model is between visioning and self-management. Self-management is the ability to use awareness of emotions to stay flexible and positively direct one's own behaviour. It is about managing emotional reactions to fit all situations and people (Bradberry & Greaves, 2001–2010). Goal setting is considered one of the important keys for developing self-management (Goleman, 2001; Manz & Sims , 1980) because of its strong reinforcing properties, subsequently leading to further goals in the pursuit of personal and organisational objectives, thus resulting in improved performance (Manz & Sims, 1980). Project management personnel should, therefore, use their visioning skill to set goals which promote their self-management development. Visioning manifested in the formulation and pursuance of personal goals is also a component that influences the activation of political skill (represented by the social astuteness component). Project management personnel should have the willingness to expend energy in pursuing personal goals. These goals and the need for achievement will push project management personnel to develop their political skill and use it to understand and build relationships with power holders to obtain valued resources paramount for goal attainment (Ferris et al., 2007).

Self-awareness, the second initiator, helps individuals understand their strengths and limitations. This understanding of oneself encourages individuals to seek feedback, learn from their mistakes, and realise when to work with others who have complementary strengths (Janasz et al., 2006). Self-awareness leads to the development of self-confidence, a significant predictor of performance (Goleman, 2001), and a key to succeed and work effectively with others (Janasz et al., 2006). The model also shows that self-awareness is a predictor of self-management, social awareness, and relationship management (all are components of emotional intelligence). These relationships have been discussed in the human skill and safety management section. It suffices to say here that self-awareness is the core of emotional intelligence (Jordan & Ashkanasy, 2006).

The third initiator, apparent sincerity, is needed to create a positive first impression, which is important in every socialisation process. First impressions are crucial in any relationship because they are lasting and influence the way people see subsequent data about the perceived object or person (Chung & Megginson, 1981). Therefore, it is important for project management personnel to appear sincere and create a first impression that they are genuinely interested in safety. In addition, apparent sincerity is needed to communicate safety vision effectively because project management personnel need to first cultivate trust before influencing people to accept the vision (Maxwell, 1993). As a consequence of this, other project stakeholders will recognise the importance of safety and project management personnel will have a positive start on influencing the others to uphold safety rules and regulations.

Apparent sincerity is also a predictor of scoping and integration (a component of conceptual skill), social awareness (a component of emotional intelligence), and social astuteness (a component of political skill). Scoping and integration is about understanding the scope of the project and integrating various project components to achieve project objectives. Apparent sincerity is important for the integration process, particularly when dealing with various project stakeholders. These stakeholders must perceive that there are no concealed motives behind any behaviour exhibited. As a result, project management personnel can inspire trust and confidence as well as influence others successfully (Ferris et al., 2005a). The influence of apparent sincerity on social awareness and social astuteness is obvious. In general, individuals with high social awareness and social astuteness have good understanding of others, thus they know how to attune themselves to diverse social situations. Apparent sincerity allows project management personnel to create favourable first impressions, which enable them to exert influence in a way that does not create ill-will or undue influence, thus other people do not question their motives or react negatively (Ferris et al., 2005b).

An interesting finding worth discussing is the exclusion of technical skill components in the model. This supports the existing theory in the management literature (Ferris et al., 2005b; Katz, 1974; Robbins et al., 2012; Samson & Daft, 2009) concerning the relationship of skills to management level as shown in Figure 4.8. Technical skill is most important for non-managerial personnel, but the need for technical skill becomes less important in higher management levels. On the other hand, conceptual skill becomes more essential as employees move up the hierarchy. Human and political skills, on the other hand, are essential at every management level. This may be due to the fact that project management personnel are not directly involved in the details of on-site activities, but instead, they are more focused on the big and overall picture of the project, in which the conceptual skill becomes more important. Furthermore, project managers normally get the construction work done through other people, such as the subcontractors. This highlights the need for human and political skills. Having said the above, we contend that technical skill should still be seen as an important skill for project management personnel as an expert-leader is more effective. In the case of something going wrong on-site, such as an accident, a formwork collapse or an improper method of work, they will need technical skill to solve the problem.

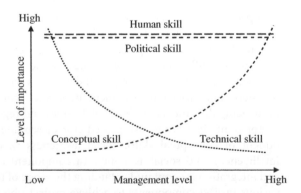

Figure 4.8   Relationship between skills and management level

Skill development strategies

The aim of a human resource development programme is to enable individuals to develop their abilities and strengths to the fullest extent. The active participation, encouragement and guidance from the superior and effective training programmes are certainly needed for development efforts to be fully productive. However, it is important to remember that no one can motivate a person towards self-development if the person has no desire to develop. Motivation must come from within, that is, the desire to learn must be intrinsic. Drucker (2008) states that development is always 'self-development'. For an organisation to assume responsibility for the development of employees is an idle boast. The responsibility rests with the individuals, their abilities and their efforts. From this motivation of self-development, a person will also realise the need to develop others. It is in and through efforts to develop others that individuals raise demands on themselves. The best performers in any profession always look upon the individuals they have trained and developed as the proudest monument they can leave behind (Drucker, 2008).

We interviewed eight industry practitioners who have, on average, more than 20 years of work experience in the construction industry to verify the skill development model. They confirmed the practicality of the model and highlighted the importance of self-awareness, visioning, apparent sincerity, social awareness and social astuteness. Therefore, the following skill development strategies will particularly focus on these skill components.

Self-awareness

In simple words, to be self-aware in the context of emotional intelligence is to know one's own emotion accurately. It is not merely about knowing preferences, but knowing oneself inside out and becoming more and more comfortable with the true essence of oneself. Not noticing and understanding emotions adequately one will find they resurface at unexpected moments and damage relationships (Bradberry & Greaves, 2009). Suggested methods for developing self-awareness are:

- Keeping records of events (people or situations) that trigger strong emotions and the results or responses of those emotions for a sufficient period of time, for example, 1 month. These records should include the physical sensations, for example, heartbeat, muscle tightness, heat or cold, that accompany the emotions. The purpose of this exercise is to see oneself more clearly and objectively by looking for emotional patterns (Bradberry & Greaves, 2009; Caruso & Salovey, 2004; Cherniss & Goleman, 2001; Mersino, 2007).
- Observing and understanding the impact of emotions on other people. Emotions can be powerful weapons which may have prolonged and detrimental effects. It is important, therefore, to immediately observe the

ripple effects of emotions on others and how they affect a wider circle long after the emotions are unleashed. This requires some time to reflect on behaviour that manifests certain emotions and also involves asking other people's opinions on how the emotions affect them (Bradberry & Greaves, 2009). Seeking feedback from others is crucial to minimise bias and get different perspectives on certain emotions and behavioural reactions. In the end, individuals should not suppress 'unknown' emotions or be afraid of emotional 'mistakes', but should learn from them (Boyatzis, 2001; Bradberry & Greaves, 2009; McCarthy & Garaven, 1999).

- Identifying the source of frustration. Everyone has a button that causes him or her to flip with anger and frustration. Knowing the people and contexts which push this button is useful to manage a person's reactions to them (Bradberry & Greaves, 2009). This exercise can be considered as backtracking, in which individuals retrace their steps to determine why they are feeling certain emotions, thus putting them in a position to do something about it (Mersino, 2007). The first three methods were designed to understand and control emotions, which can be particularly useful when delivering safety-related talks. A strong but appropriate expression of emotions can enhance the impact of these safety talks on others.
- Developing an objective understanding of behaviour. This involves taking notice of and reflecting on emotions, thoughts and behaviours immediately when the situation unfolds. Such objectivity allows individuals not to be controlled by emotions and moods, thus knowing exactly what to be done to create a positive outcome (Bradberry & Greaves, 2009; Cherniss & Goleman, 2001; Mersino, 2007). For example, rather than screaming and being angry with workers who do not follow safety procedures, a safety officer can consider and reflect on the sources of those unsafe behaviours.
- Personality tests, such as the Myers–Briggs Type Indicator (MBTI), can also be used to develop self-awareness (Cherniss & Goleman, 2001; McCarthy & Garaven, 1999). The MBTI is used to explain the effects of personal preferences on decision making and problem solving by looking at eight behavioural preferences that all people use to some degree. The instrument is useful in helping people become more self-aware of their preferences and characters (McCarthy & Garaven, 1999). A study has revealed that there is a strong association between personality types and number of injuries, demonstrating the relevance of this self-awareness development method in improving safety performance (Pierce, 2005).

The self-awareness development methods mentioned above aim to improve the way individuals manage themselves and their interpersonal relationships, which indirectly influence safety. There are also methods that can directly impact on safety performance. In the area of transportation safety, the Driving Decisions Workbook has been developed to give individuals a source of information about themselves with regard to their current or future driving. A study has shown that the workbook is effective in increasing self-awareness in relation to safe driving (Eby et al., 2003). Another research study, also in transportation safety, proposes the use of self-screening to increase self-awareness about functional abilities associated with safe driving. The self-screening instrument consists of 27 health concerns and 15 critical driving skills, resulting in increased awareness of how functional declines can affect driving performance (Molnar et al., 2010).

A similar approach may be applicable in the construction industry. A self-assessment workbook or checklist can be used by workers and staff to assess themselves to ensure that they are aware of and understand basic safety requirements and procedures before they enter construction sites or perform certain tasks. In addition, a self-screening instrument can also be used to assess whether an individual is in a proper condition and qualified to work on-site.

Visioning

Intuition, critical thinking and decisiveness are three underlying aspects of visioning. Intuition is the knowledge gained through discrimination or discernment where a person's mind directly discerns or comprehends cause–effect relationships. An intuitive person comprehends which events really shape the future, while being flexible enough to admit critical change elements as they occur, thus enabling the person to shake free of the limitation of time and past conditioning, and create an accurate vision of the best course of action (Nuernberger, 1992).

Critical thinking is the ability to synthesise and analyse data, to perceive clearly and to discriminate. The power of discrimination enables a person to think critically by discerning critical elements from a mass of data, being flexible in using information that appears inconsistent or irrelevant and perceiving the entire picture. The capacity for critical thinking helps a person solve problems in bringing about a vision of the future (Nuernberger, 1992). This vision at the same time should be an ideal to strive for and created from both intuition (right-brain thinking) and logical analysis (left-brain thinking) (Tichy & Devanna, 1990).

Decisiveness is about knowing when and how to take action. Creative solutions, brilliant ideas and intuitive insights have little value unless they are converted into action. A decisive person knows the time to take action when the opportunity presents itself. This action conserves effort and ensures success, while 'he who hesitates is lost'. The real reasons for indecision and inability to act are fear and self-doubt. Plagued with doubt and the fear of making mistakes, many people hesitate and fail to decide in a timely fashion. Therefore, it is necessary to learn to take fear out of the decision-making process and bypass self-doubt by becoming an experimenter and adventurer (Nuernberger, 1992).

When it comes to implementation, the very first step in visioning is to create an effective vision relevant to the organisation or the project. There are three components of a compelling and believable vision. The first component is continuity because an effective vision must carry forward the best of the past. This ensures that the past contributions to it are valued, thus allowing people to experience an essential sense of continuity and worth as they create anew.

The second component is innovation. It must extend the old into the new and offer a challenge worthy of the best efforts of those who will create it, while also connecting with and expressing the dreams, hopes and aspirations of the individuals who make up the team (Vogt, 2009). In this case, a vision should have an emotional appeal element, having a motivational pull with which people can identify (Tichy & Devanna, 1990). The third component is transition, meaning that it must be achievable. People must be engaged not only in its creation, but also in its implementation. In principle, a complete and compelling vision is a vision that builds on the past, extends into the future and suggests a bridge to that future (Vogt, 2009). It provides a conceptual framework for understanding the team's purpose and includes a roadmap (Tichy & Devanna, 1990). These three components should be taken into consideration in formulating the safety vision that senior managers and project management personnel intend to promote in the organisation.

Grounded visioning is a practical method that can be used to help project management personnel or senior managers discover their shared safety visions within a short period of time. Grounded visioning focuses on generating visionary ideas that are grounded in past achievements. There are six steps to developing a grounded vision (Vogt, 2009):

1. Bringing together key stakeholders who have a stake in the safety implementation in the organisation or project.
2. Igniting passion for what is possible.
3. Sharing best safety practices by asking the key stakeholders to call out what attracted them to the organisation or project and what keeps them involved in safety implementation.
4. Sharing safety dreams for the future and identifying emerging themes.
5. Reaching consensus by identifying the top safety vision(s).
6. Determining ways to bring the vision into reality.

In relation to the last step above, a vision needs to be translated into goals at different levels of management. The criteria for effective goals are (Samson & Daft, 2009):

1. *Specific and measurable.* When possible, goals should be expressed in quantitative terms, for example, zero accident, 20% reduction of first-aid injuries. The rule is that goals must be precisely defined and allow for measurable progress because vague goals have little motivating power.
2. *Covering key result areas.* It is impossible to set goals and measure every aspect of safety behaviour and performance, as the sheer number of these goals will render them meaningless. Therefore, key result areas covering activities that contribute most to safety performance have to be identified.
3. *Challenging but realistic.* If goals are too easy, people may not feel motivated. On the contrary, when goals are unrealistic, they set people up for failure and lead to poor morale. Safety should be considered as a journey and not a goal or commitment that can be developed overnight.

4. *Defined period.* Goals should specify the period over which they will be achieved, which includes a deadline stating the date on which goal attainment will be measured.

5. *Linked to rewards.* The ultimate impact of goals depends on the extent to which salary increases, promotions, and awards are based on goal achievement. Rewards give meaning and significance to goals and help people commit to achieve the goals.

Visioning, as part of conceptual skill, is very abstract, making it difficult to be developed by using conventional methods. Conceptual skill (including visioning) is strongly correlated with years of work experience, thus indicating that on-the-job experience through interactions with various stakeholders and project components may be the best 'method' to develop this skill. In the end, visioning skill involves lifelong learning and should be refined continuously. It is about reflection and looking within oneself to recognise values, principles, dreams, feelings and even 'higher calling' that lifts a person above one's self. It is about observing what is happening to others and identifying resources that are available. Ultimately and more importantly, it is about learning from past experiences (Maxwell, 1993). It is important to recognise that experience can be the best teacher. Organisations and managers, therefore, should focus on the way they and others can learn from mistakes, rather than fostering a climate in which people hide mistakes because of fear of punishment (Samson & Daft, 2009; Sunindijo & Zou, 2013).

Apparent sincerity

This skill is key in influencing project stakeholders to be committed towards safety and follow safety procedures. Drama-based training and behavioural modelling techniques are methods suggested to develop this skill by demonstrating and observing different ways of showing sincerity or insincerity. Actors could role-play various levels of sincerity for different situations and then these genuineness variations are practised by trainees. It should be noted that appearing authentic is not only a matter of words, but also of tone, silences, gestures, facial expression and deep feeling for the purposes that are being presented (Ferris et al., 2005b).

Charismatic communication is another effective way to present the importance of certain values by conveying the desired impression (Ferris et al., 2005b). People who have received charismatic communication training employ more animated gestures, analogies and stories, and were perceived as effective communicators (Frese et al., 2003). In the context of safety management, charismatic leaders are able to communicate safety vision that inspires others. Action training has been proved as a method that can be used to improve charismatic communication, which is valuable to communicate an inspirational safety vision. Frese et al. (2003) proposed six components of action training. The six components below have been contextualised to demonstrate how safety can be incorporated into an action training programme:

1. *Action-oriented mental model of what constitutes effective actions within certain situations.* This model is a cognitive representation of the starting situation, the goal state, and how the present situation can be transferred into a future state. It provides a set of principles or learning points to describe charismatic and inspirational communication.
2. *Learning by doing, which encourages trainees to take an active approach and learn by means of role-play procedures.* Trainees can be asked to do a role-play in communicating safety vision inspirationally to their employees several times to ensure that the general principles are associated to their actions.
3. *Motivation by experiencing or visualising the difference between the present state and future safety goals.* The intention is to show that certain goals are yet to be achieved. Trainees are encouraged to experiment, make mistakes, and then learn from their own mistakes and those of other trainees. This type of motivation is consistent with goal setting theory asserting that challenging and rewarding goals produce high motivation.
4. *Feedback in training by the trainer, other participants, and one's own critical evaluation.* The feedback should be given with a functional task perspective, that is, relating the feedback to the task to reduce the artificial nature of the feedback.
5. *Supporting transfer of the principles that have been learnt into a work situation.* This is crucial to ensure that trainees develop knowledge by applying the principles to their work. In addition, trainees are asked to think of specific dates when they will use their newly developed skills.
6. *The necessity to routinise behaviour.* Newly developed skills typically will compete with old, well-rehearsed routines. Therefore, it is necessary to repeat performance during training to create a certain amount of routinisation and discuss practical issues right after the training.

Social awareness and social astuteness

Both social awareness and social astuteness skill components are essentially about understanding people. Through this understanding of people, proper tactics and approaches can be determined to influence project stakeholders to focus on safety. In order to develop these skills, Cherniss and Goleman (2001) suggested training that promotes empathy. For example, trainees are shown pictures of actors expressing different emotions and they try to understand what emotions the actors are expressing. Gradually, different parts of each actor's face are obscured to make the task harder. Similarly, Caruso and Salovey (2004) suggested that social awareness can be improved by paying attention to and actively looking for emotional clues. Three main sources of emotional clues that help individuals identify others' emotions accurately are facial expression; the pitch, rhythm, and tone of voices; and the feelings conveyed by the posture of someone's body.

Mersino (2007) argued that listening and observing are important in developing social awareness. The self-tracking emotion journal used to develop self-awareness can also be expanded to include the emotions of others. Another development method to be considered is observing and learning from people who are effective at social awareness and social astuteness. In addition, drama-based training and role playing are recommended to help trainees learn the more astute ways of understanding, and interacting with, others (Ferris et al., 2005b). Finally, it is also necessary to observe the safety culture in the project and assess the stakeholders to contextualise the learning environment, thus individuals can adjust themselves based on the environment (Mersino, 2007).

Dimitrius and Mazzarella (1999) suggested similar methods to understand people and predict their behaviour. They argued that understanding people involves 'hearing more than just words', signifying the need to identify hidden meanings behind words through physical appearance, body language and intuition. This is particularly true and useful in the multi-cultural construction industry, like those in Australia, the USA, UK and Dubai. Observing a person's environment, for example, appearance, clothing, house and room, can also reveal clues about that person and confirming or deepening what the observer has noticed from the personal appearance and body language. However, it is important to understand some exceptions to the rules, which if not taken into account, may create distorted or incorrect conclusions, although everything else seems to fit.

Conclusions

Project management personnel have important roles in developing, implementing and evaluating safety management strategies in construction and engineering projects. This chapter has discussed four essential skills (and their components), namely, conceptual skill, human skill, political skill and technical skill, in order for project management personnel to play their safety roles successfully in construction and engineering projects. A tiered model was developed to demonstrate the interrelationships between the important skill-for-safety components. The foundational skills are self-awareness, visioning and apparent sincerity. The first tier mediator skills are scoping and integration, and self-management. The second tier mediator skills are social awareness, social astuteness and relationship management. These skills affect the third tier mediator (the implementation of safety management tasks), leading to the ultimate outcome which is safer construction. Strategies for developing these skills have also been proposed. It should be noted that technical skill components have been omitted from the model. It does not mean that they are not important, but the higher up the management ladder, the greater the emphasis should be on the other three skills rather than on technical proficiency.

Managers, both at the organisational and project levels, should be aware of the influence of these skills on safety management, particularly in the construction context. In designing their human resource development programme, they need to help project management personnel assess and prioritise their skill development, which will enable them to lead the implementation of safety strategies. The skill development model provides simplified relationships between the skill components and suggests how project management personnel should develop their skills for safety. The model should help senior managers and project management personnel simplify and make sense of the complex social process involved in the skill development and application to influence the implementation of safety strategies. It should be noted, however, that the model should neither be viewed as rigid nor linear in nature. In reality, these skills should cohere and function simultaneously to produce the desired impacts on improving safety performance in construction and engineering projects; thus it is almost impossible to establish boundaries and determine when a certain skill functions in social interactions. Furthermore, project management personnel should understand their own strengths and weaknesses so that their skill development process can be tailored according to their needs and existing skill sets.

References

Abudayyeh, O., Fredericks, T. K., Butt, S. E., & Shaar, A. (2006). An investigation of management's commitment to construction safety. *International Journal of Project Management, 24*(2), 167–174.

Ahearn, K. K., Ferris, G. R., Hochwarter, W. A., Douglas, C., & Ammeter, A. P. (2004). Leader political skill and team performance. *Journal of Management, 30*(3), 309–327.

Akintoye, A. S., & MacLeod, M. J. (1997). Risk analysis and management in construction. *International Journal of Project Management, 15*(1), 31–38.

Aksorn, T., & Hadikusumo, B. H. W. (2008). Critical success factors influencing safety program performance in Thai construction projects. *Safety Science, 46*(4), 709–727.

Anton, T. J. (1989). *Occupational Safety and Health Management* (2nd ed.). New York: McGraw-Hill.

Australian Standards & New Zealand Standards. (2009). *AS/NZS ISO 31000:2009 Risk Management - Principles and Guidelines.* Sydney, Australia and Wellington, New Zealand: Standards Australia/Standards New Zealand.

Bass, B. M., & Riggio, R. E. (2006). *Transformational Leadership* (2nd ed.). Mahwah, NJ: Lawrence Erlbaum Associates.

Blickle, G., Kramer, J., Zettler, I., Momm, T., Summers, J. K., Munyon, T. P., & Ferris, G. R. (2009). Job demands as a moderator of the political skill-job performance relationship. *Career Development International, 14*(4), 333–350.

Blickle, G., Kramer, J., Schneider, P. B., Meurs, J. A., Ferris, G. R., Mierke, J., Momm, T. (2011). Role of political skill in job performance prediction beyond general mental ability and personality in cross-sectional and predictive studies. *Journal of Applied Social Psychology, 41*(2), 488–514.

Boyatzis, R. E. (2001). How and why individuals are able to develop emotional intelligence. In C. Cherniss & D. Goleman (Eds.), *The Emotionally Intelligence Workplace* (pp. 234–253). San Francisco, CA: Jossey-Bass.

Bradberry, T., & Greaves, J. (2001–2010). *The Emotional Intelligence Appraisal - Me Edition: There is more than IQ*. San Diego: TalentSmart.

Bradberry, T., & Greaves, J. (2009). *Emotional Intelligence 2.0*. San Diego, CA: TalentSmart.

Brill, J. M., Bishop, M. J., & Walker, A. E. (2006). The competencies and characteristics required of an effective project manager: A web-based Delphi study. *Educational Technology, Research and Development, 54*(2), 115–140.

Brouer, R. L., Duke, A., Treadway, D. C., & Ferris, G. R. (2009). The moderating effect of political skill on the demographic dissimilarity - Leader-member exchange quality relationship. *The Leadership Quarterly, 20*(2), 61–69.

Buchanan, D., & Badham, R. (1999). *Power, Politics, and Organizational Change: Winning the Turf Game*. London: Sage Publications.

Byrne, B. M. (2010). *Structural Equation Modeling with AMOS Basic Concepts, Applications, and Programming* (2nd ed.). New York: Routledge.

Carter, G., & Smith, S. D. (2006). Safety hazard identification on construction projects. *Journal of Construction Engineering and Management, 132*(2), 197–205.

Caruso, D. R., & Salovey, P. (2004). *The Emotionally Intelligent Manager: How to Develop and Use the Four Key Emotional Skills of Leadership*. San Francisco, CA: Jossey-Bass.

Chen, P., Partington, D., & Wang, J. N. (2008). Conceptual determinants of construction project management competence: A Chinese perspective. *International Journal of Project Management, 26*(6), 655–664.

Cheng, M., Dainty, A. R. J., & Moore, D. R. (2005). What makes a good project manager? *Human Resource Management Journal, 15*(1), 25–37.

Cherniss, C., & Goleman, D. (2001). Training for emotional intelligence: A model. In C. Cherniss & D. Goleman (Eds.), *The Emotionally Intelligent Workplace* (pp. 209–233). San Francisco, CA: Jossey-Bass.

Chung, K. H., & Megginson, L. C. (1981). *Organizational Behavior: Developing Managerial Skills*. New York: Harper & Row.

Dainty, A. R. J., Cheng, M., & Moore, D. R. (2003). Redefining performance measures for construction project managers: an empirical evaluation. *Construction Management and Economics, 21*(2), 209–218.

Damasio, A. (1994). *Descartes' Error: Emotion, Reason, and the Human Brain*. New York: Grosset/Putnam.

Dimitrius, J.-E., & Mazzarella, M. (1999). *Reading People: How to Understand People and Predict Their Behavior, Anytime, Anyplace*. New York: Ballantine Books.

Dingsdag, D. P., Biggs, H. C., Sheahan, V. L., & Cipolla, D. J. (2006). *A Construction Safety Competency Framework: Improving OH&S Performance by Creating and Maintaining a Safety Culture*. Brisbane: Cooperative Research Centre for Construction Innovation.

Drucker, P. (1996). Foreword. In F. Hesselbein, M. Goldsmith & R. Beckhard (Eds.), *The Leader of the Future: New Visions, Strategies and Practices for the Next Era* (pp. xi). San Francisco: Jossey-Bass.

Drucker, P. (2008). *Management* (Revised ed.). Pymble, NSW, Australia: HarperCollins.

Eby, D. W., Molnar, L. J., Shope, J. T., Vivoda, J. M., & Fordyce, T. A. (2003). Improving older driver knowledge and self-awareness through self-assessment: The driving decisions workbook. *Journal of Safety Research, 34*(4), 371–381.

Ellis, C. W. (2005). *Management Skills for New Managers*. New York: Amacom.

El-Sabaa, S. (2001). The skills and career path of an effective project manager. *International Journal of Project Management, 19*(1), 1–7.

Farooqui, R. U., Saqib, M., & Ahmed, S. M. (2008). Assessment of critical skills for project managers in Pakistani construction industry. Paper presented at the First International Conference on Construction in Developing Countries (ICCIDC-I), Karachi, Pakistan.

Ferris, G. R., Perrewé, P. L., Anthony, W. P., & Gilmore, D. C. (2000). Political skill at work. *Organizational Dynamics, 28*(4), 25–37.

Ferris, G. R., Davidson, S. L., & Perrewé, P. L. (2005a). Developing political skill at work. *Training, 42*(11), 40–45.

Ferris, G. R., Perrewé, P. L., Anthony, W. P., & Gilmore, D. C. (2005b). *Political Skill at Work: Impact on Work Effectiveness.* Mountain View, CA: Davies-Black Publishing.

Ferris, G. R., Treadway, D. C., Kolodinsky, R. W., Hochwarter, W. A., Kacmar, C. J., Douglas, C., & Frink, D. D. (2005c). Development and validation of the political skill inventory. *Journal of Management, 31*(1), 126–152.

Ferris, G. R., Treadway, D. C., Perrewé, P. L., Brouer, R. L., Douglas, C., & Lux, S. (2007). Political skill in organizations. *Journal of Management, 33*(3), 290–320.

Fisk, E. R. (1997). *Construction Project Administration* (5th ed.). Upper Saddle River, NJ: Prentice Hall.

Frese, M., Beimel, S., & Schoenborn, S. (2003). Action training for charismatic leadership: Two evaluations of studies of a commercial training module on inspirational communication of a vision. *Personnel Psychology, 56*(3), 671–697.

Gillard, S., & Price, J. (2005). The competencies of effective project managers: A conceptual analysis. *International Journal of Management, 22*(1), 48–53.

Goleman, D. (1998). *Working with Emotional Intelligence.* New York: Bantam Books.

Goleman, D. (2001). An EI-based theory of performance. In C. Cherniss & D. Goleman (Eds.), *The Emotionally Intelligent Workplace* (pp. 27–44). San Francisco: Jossey-Bass.

Goodwin, R. S. C. (1993). Skills required of effective project managers. *Journal of Management in Engineering, 9*(3), 217–226.

Gushgari, S. K., Francis, P. A., & Saklou, J. H. (1997). Skills critical to long-term profitability of engineering firms. *Journal of Management in Engineering, 13*(2), 46–56.

Hayes, J. (1984). The politically competent manager. *Journal of General Management, 10*(1), 24–33.

Holden, M. (1998). *Positive Politics: Overcome Office Politics & Fast-track Your Career.* Warriewood, NSW, Australia: Business & Professional Publishing.

Hudson, P. (2007). Implementing a safety culture in a major multi-national. *Safety Science, 45*(6), 697–722. doi: 10.1016/j.ssci.2007.04.005

Janasz, S. D., Wood, G., Gottschalk, L., Dowd, K., & Schneider, B. (2006). *Interpersonal Skills in Organisations.* Boston: McGraw-Hill.

Jordan, P. J., & Ashkanasy, N. M. (2006). Emotional intelligence, emotional self-awareness, and team effectiveness. In V. U. Druskat, F. Sala & G. Mount (Eds.), *Linking Emotional Intelligence and Performance at Work: Current Research Evidence with Individuals and Groups* (pp. 145–163). New Jersey: Lawrence Erlbaum Associates.

Katz, R. L. (1974). Skills of an effective administrator. *Harvard Business Review,* Sep–Oct, 90–102.

Kerzner, H. (2009). *Project Management: A System Approach to Planning, Scheduling, and Controlling* (10th ed.). Hoboken, NJ: Wiley.

Koegel, L. K., Koegel, R. L., Hurley, C., & Frea, W. D. (1992). Improving social skills and disruptive behavior in children with autism through self-management. *Journal of Applied Behavior Analysis, 25*(2), 341–353.

Lane, R. D. (2000). Levels of emotional awareness: neurological, psychological, and social perspectives. In R. Bar-On & J. D. A. Parker (Eds.), *The Handbook of Emotional Intelligence* (pp. 171–191). San Francisco: Jossey-Bass.

Langford, D., Rowlinson, S., & Sawacha, E. (2000). Safety behaviour and safety management: its influence on the attitudes of workers in the UK construction industry. *Engineering, Construction and Architectural Management, 7*(2), 133–140.

Lei, W. W. S., & Skitmore, M. (2004). Project management competencies: A survey of perspectives from project managers in South East Queensland. *Journal of Building and Construction Management, 9*(1), 1–12.

Lientz, B. P., & Rea, K. P. (2002). *Project Management for the 21st Century* (3rd ed.). San Diego: Butterworth Heinemann.

Mainiero, L. A. (1994). On breaking the glass ceiling: The political seasoning of powerful women executives. *Organizational Dynamics, 22*(4), 5–20.

Manz, C. C., & Sims Jr., H. P. (1980). Self-management as a substitute for leadership: A social learning theory perspective. *Academy of Management Review, 5*(3), 361–367.

Mattila, M., Hyttinen, M., & Rantanen, E. (1994). Effective supervisory behaviour and safety at the building site. *International Journal of Industrial Ergonomics, 13*(2), 85–93.

Maxwell, J. C. (1993). *Developing the Leader Within You.* Nashville: Thomas Nelson.

McCarthy, A. M., & Garaven, T. N. (1999). Developing self-awareness in the managerial career development process: The value of 360-degree feedback and the MBTI. *Journal of European Industrial Training, 23*(9), 437–445.

Mersino, A. C. (2007). *Emotional Intelligence for Project Managers: The People Skills You Need to Achieve Outstanding Results.* New York: AMACOM.

Molnar, L. J., Eby, D. W., Kartje, P. S., & St. Louis, R. M. (2010). Increasing self-awareness among older drivers: The role of self-screening. *Journal of Safety Research, 41*(4), 367–373.

Nuernberger, P. (1992). *Increasing Executive Productivity: A Unique Program for Developing the Inner Skills of Vision, Leadership, and Performance.* Englewood Cliffs, New Jersey: Prentice Hall.

Odusami, K. T. (2002). Perceptions of construction professionals concerning important skills of effective project leaders. *Journal of Management in Engineering, 18*(2), 61–67.

Peled, A. (2000). Politicking for success: the missing skill. *The Leadership and Organization Development Journal, 21*(1), 20–29.

Perrewé, P. L., & Nelson, D. L. (2004). Gender and career success: the facilitative role of political skill. *Organizational Dynamics, 33*(4), 366–378.

Peterson, T. O., & Fleet, D. D. V. (2004). The ongoing legacy of R.L. Katz: An updated typology of management skills. *Management Decision, 42*(10), 1297–1308.

Pierce, F. D. (2005). Personality types and injuries: A statistical study and effective strategies. *Professional Safety, 50*(3), 42–50.

Pinto, J. K. (1998). Power, politics, and project management. In J. K. Pinto (Ed.), *The Project Management Institute: Project Management Handbook* (pp. 256–266). San Francisco: Jossey-Bass.

Pinto, J. K. (2000). Understanding the role of politics in successful project management. *International Journal of Project Management, 18*(2), 85–91.

Project Management Institute. (2013). *A Guide to the Project Management Body of Knowledge: PMBOK Guide* (5th ed.). Pennsylvania, USA: Project Management Institute.

Robbins, S. P., & Hunsaker, P. L. (2009). *Training in Interpersonal Skills: TIPS for Managing People at Work*. Upper Saddle River, NJ: Pearson Prentice Hall.

Robbins, S. P., Bergman, R., Stagg, I., & Coulter, M. (2012). *Management* (6th ed.) Australia: Pearson.

Samson, D., & Daft, R. L. (2009). *Management* (3rd Asia Pacific ed. South Melbourne: Cengage Learning Australia.

Smith, A. D., Plowman, D. A., Duchon, D., & Quinn, A. M. (2009). A qualitative study of high-reputation plant managers: Political skill and successful outcomes. *Journal of Operations Management, 27*(6), 428–443.

Spencer, L. M., & Spencer, S. M. (1993). *Competence at Work: Models for Superior Performance*. New York: John Wiley & Sons.

Sunindijo, R. Y., & Zou, P. X. W. (2011). CHPT construct: Essential skills for construction project managers. *International Journal of Project Organisation and Management, 3*(2), 139–163.

Sunindijo, R. Y., & Zou, P. X. W. (2012a). How project manager's skills may influence the development of safety climate in construction projects. *International Journal of Project Organisation and Management, 4*(3), 286–301.

Sunindijo, R. Y., & Zou, P. X. W. (2012b). Political skill for developing construction safety climate. *Journal of Construction Engineering and Management, 138*(5), 605–612.

Sunindijo, R. Y., & Zou, P. X. W. (2013). Conceptualizing safety management in construction projects. *Journal of Construction Engineering and Management, 139*(9), 1144–1153.

Sunindijo, R. Y., & Zou, P. X. W. (2014). The roles of emotional intelligence, interpersonal skill, and transformational leadership on improving construction safety performance. *Australasian Journal of Construction Economics and Building, 13*(3), 97–113.

Tichy, N. M., & Devanna, M. A. (1990). *The Transformational Leader*. New York: John Wiley & Sons.

Vigoda, E. (2003). *Developments in Organizational Politics: How Political Dynamics Affect Employee Performance in Modern Worksites*. Cheltenham, UK: Edward Edgar.

Vogt, J. W. (2009). *Recharge Your Team: The Grounded Visioning Approach*. Westport: Praeger.

Wang, W. C., Liu, J. J., & Chou, S. C. (2006). Simulation-based safety evaluation model integrated with network schedule. *Automation in Construction, 15*(3), 341–354.

Zou, P. X. W., & Sunindijo, R. Y. (2010). Construction safety culture: A revised framework. Paper presented at the The Chinese Research Institute of Construction Management (CRIOCM), 15th annual symposium, Johor Bahru, Malaysia.

Zou, P. X. W., & Sunindijo, R. Y. (2013). Skills for managing construction safety risks, implementing safety tasks and developing positive safety climate. *Automation in Construction, 34*, 92–100.

5 Safety Training and Learning

In this chapter, we discuss safety training and learning. We refer to safety training as the programmes and processes imposed externally by the construction regulatory bodies, the industry and the organisation, whereas safety learning is focused on how trainees themselves learn in training programmes, in work places and via other self-learning opportunities. We also discuss the role of the trainer and training pedagogies. The construction industry is characterised by temporary organisations and extensive outsourcing in which workers often move from one job to another within a relatively short period of time. Therefore, it is important that workers attain safety knowledge and have a strong awareness of safe and correct practices, in order to minimise safety risks in the dynamic environment that characterises their work place (Wilkins, 2011). Furthermore, safety learning is important not only to provide safety knowledge and safety skills, which have been discussed in Chapter 4, but also serves as a way to socialise an organisation's safety culture, which was discussed in Chapter 3.

This chapter starts with a general discussion on the nature of pedagogy and andragogy in order to show that formal safety training programmes can be delivered using a combination of these two approaches to improve the effectiveness of training and learning. This is followed by a discussion on common issues in safety learning in the construction industry and offers recommendations to address those issues. It is also argued that safety learning does not predominantly occur through formal training channels, but informally through on-the-job practices and interactions with artefacts and people. The chapter concludes by presenting techniques to evaluate the effectiveness of safety training programmes, along with a case study.

Strategic Safety Management in Construction and Engineering, First Edition.
Patrick X.W. Zou and Riza Yosia Sunindijo.
© 2015 John Wiley & Sons, Ltd. Published 2015 by John Wiley & Sons, Ltd.

Training and learning defined

Employee training can be defined as a systematic effort (often initiated by an employer) to develop the knowledge, skills and attitudes required by an employee to perform a given task or job successfully with the final goal to improve the organisation's performance (Kalliath et al., 2010). Training is part of human resource development, which aims to develop employees' full potential irrespective of the possibility of its immediate use in the current job. Such a focus on human resource development can be attributed to the growing realisation that people are the main source of competitive advantage and that an organisation's success is determined by decisions that employees make and behaviours in which they engage (Ruona & Gibson, 2004).

Training can be differentiated from coaching and education. Coaching refers to a process of a one-on-one relationship between a coach and an employee to generate behavioural changes through self-awareness and learning (Joo, 2005). It is about focusing on the current interests and competencies of individual employees, and trying to develop them through personalised attention. Education, as a human resource development, also aims at developing individuals' potential, not only to make them more effective in a particular organisation, but to enable those individuals to manage their lives in different contexts (Kalliath et al., 2010).

There are three training paradigms underlying the provision of training in organisations: *the traditional paradigm, the human resource development paradigm* and *the learning paradigm*. The *traditional paradigm* views training as a necessity which is driven by the employer's perceptions of the competency-gap in employees when performing their current job. In the *human resource development paradigm*, although the employer's perceptions concerning the developmental needs of employees are still the driver, the scope goes not only beyond their current job but also includes a match between the organisation's future operations and the employees' future potential. The *learning paradigm* empowers employees to set and implement their learning agendas within the framework of organisational goals, values and resource constraints. The philosophy of the learning paradigm is that learning is most effective when it is initiated by the learners and that the organisation's role is to create a supportive environment for learning. It should be noted that although it is possible to distinguish between the three paradigms at the conceptual level, in practice their features overlap, depending on the situation (Kalliath et al., 2010). Although this chapter will include some elements of the traditional paradigm and the human resource development paradigm, the main aim is to assist organisations in embedding the learning paradigm as an integral part of safety training and learning practices.

Approaches to learning: pedagogy and andragogy

Organisational learning is important because of its positive impacts on productivity and performance. The term "learning curves" as shown in Figure 5.1,

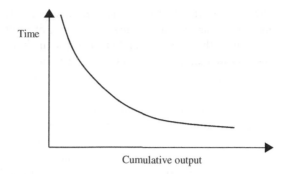

Figure 5.1   Typical learning curve

for example, is widespread in many organisations across a variety of industries, including assembly plants, factories, healthcare procedures, farms, construction and food production. The learning-curve pattern essentially shows that the time individuals take to perform a task and the number of errors they make decrease at a decreasing rate as experience is gained with the task. Essentially it shows that at first new skills and knowledge can be acquired quickly which yield significant impacts, but subsequent learning is slower with lesser impacts on the outcome (Argote, 1999).

The learning-curve concept boosted the popularity of organisational learning which has been manifested in various knowledge transfer efforts in recent years. People argue that an organisation that is able to transfer knowledge, such as productivity improvement and better quality services, made at one establishment to another will have the upper hand against those who are ineffective at knowledge transfer. Consequently, organisations have focused on improving their learning rates by increasing the proficiency of employees at all levels; improving technology; and improving organisational structure, routines and coordination methods (Argote, 1999).

There are two main approaches to learning in a formal context. The first approach dominated secular schools and emerging universities in Europe at the end of the twelfth century. This approach is called *pedagogy*, a term derived from the Greek words *paid* (meaning child) and *agogus* (meaning leading). Therefore, pedagogy literally means the art and science of teaching children. The pedagogical assumptions of learning are initially based on observations in teaching young children basic skills, mostly reading and writing. When people started to realise the importance of education, this approach was spread and adopted across the world. This approach was eventually so strongly embedded in society that, even in the early twentieth century, educational psychologists constrained their research around pedagogical assumptions by studying the reactions of children and animals to a variety of didactic teaching and instructions. When organised adult education was initiated in the 1920s, several problems relating to the pedagogical approach began to emerge. Adult learners seem to be resistant to pedagogical methods of learning, such as fact-laden lectures, assigned readings, quizzes, rote memorising and examinations. As a result, the drop-out rates among adult learners were high. It appears that many of the assumptions about the characteristics of learners do not fit adult learners (Knowles, 1980).

At about that time, researchers began to analyse successful teachers of adults and how they adopted methods which deviated from the pedagogical principles. Adult educators then coined the term *andragogy*, which is based on the Greek word *anēr*, meaning man, not boy or adult. At first, andragogy was defined as the art and science of helping adults learn. Increasing number of teachers, however, found that the andragogical approach also produced superior learning among youth. Therefore, instead of classifying pedagogy and andragogy as learning for children and learning for adults respectively, we should see them as two approaches that fit to particular situations. They represent two ends of a spectrum with a realistic assumption in a given situation falling between the two ends (Knowles, 1980). Table 5.1 summarises the difference in assumptions between pedagogy and andragogy.

Based on the above discussions, the pedagogy approach, in which trainers take complete responsibility for all decisions related to learning while trainees only play a dependent role of following the instructions of the trainers, is *not* the best way to train construction workers. Instead we argue that the andragogy approach is more suitable for construction safety training and learning. Construction workers as adult learners, particularly those with significant experience in the industry, are different from dependent traditional learners. First, they are independent and self-directing, thus they dislike being imposed on or directed by others. Second, they have accumulated a great deal of experience, which is a valuable resource for mutual learning. Third, they value learning that integrates with the demands of their everyday life, thus they need to know the significance and benefits of why they should undertake a learning experience. Fourth, they are more interested in immediate, problem-centred approaches which will assist them in dealing with practical situations they may encounter. Fifth, they are more motivated to learn by internal drivers, for example, job satisfaction and quality of life, than external ones, for example, promotions and salaries (Albert & Hallowell, 2013; Kaufman, 2003).

When dealing with non-traditional adult learners, Albert and Hallowell (2013) proposed six core components to facilitate safety training, which are based on the andragogy approach:

1. Develop a safety performance model

    This is about developing a model of desired safety performance usually achieved through brainstorming sessions to expose areas of concern with the current safety performance. An external facilitator may be used to assist in this process by introducing models from external sources. This activity allows employees to understand the benefits of acquiring safety knowledge and skills, thus motivating them to learn to improve their performance.

Table 5.1   Different assumptions of pedagogy and andragogy (Albert & Hallowell, 2013; Knowles et al., 1998)

Assumption	Pedagogy	Andragogy
Need to know	Learners do not necessarily know the benefits of learning. They follow instructions to receive good feedback and evaluation.	Learners need to know the significance and benefits of learning before they endeavour to learn.
Self-concept of learners	Learners are dependent and rely on the expertise of the instructor and accept most imposed learning methods.	Learners feel responsible for their own decisions and life. As such, they are self-directing and dislike being imposed upon or directed by others.
Experience of learners	The experience of learners is minimal, only the experience and knowledge of the instructor and the instructor-provided material are important.	Learners have different backgrounds and have gathered a considerable amount of experience which are valuable resources for mutual learning.
Readiness to learn	Learners are ready to learn anything imparted by the instructor and ready to follow instructions.	Learners are open to learn things that are essential or those that assist them in dealing with the practical situations they may encounter.
Orientation to learn	Learners are subject-centred. They follow the organised contents prepared by the instructor.	Learners are task or problem-centred. They are motivated to learn when they perceive that learning will help them deal with problems that they confront in their life situations.
Motivation to learn	Learners are motivated by external factors such as grades, instructor approval, and parental pressure.	Learners are motivated by external factors, such as promotions and higher salaries, but are driven more by internal factors such as job satisfaction, self-esteem and quality of life.

2. Develop appraisal of safety competence

In this step, employees should benchmark their current practice with the desired level of safety performance as established in the previous step. This assessment would allow employees to identify performance gap where learning is required. The facilitator may assist in this process so that employees are able to assess themselves more objectively.

3. Devise safety learning objectives

    Based on the performance gap identified in the previous step, learning plans are developed to improve competency and performance. Employees should be actively involved in creating these plans and their learning objectives, which are based on the assumption that adult learners are self-directing.

4. Develop and implement safety learning strategies

    This step is about the identification of learning resources and the development of learning strategies to meet the developmental needs to close the performance gap. A wide range of human and material resources should be made available to employees throughout the learning process.

5. Measure and assess safety learning outcomes

    This step aims to facilitate continuous improvement of safety learning. There are four factors that should be assessed: response from the trainees to the learning process; knowledge and skills gained; behaviour changes induced as a result of learning; and benefits received by the organisation through the improvement in performance. These four factors will be discussed in detail in the 'safety learning evaluation techniques' section later in this chapter.

6. Foster a climate and environment for safety learning

    This refers to an environment conducive for successful integration of the previous five steps into the business processes. This environment includes both the psychological climate, such as mutual respect, trust, openness and management commitment, and to physical resources, such as the availability of facilitators, learning materials and other physical environments that support learning. Furthermore, it is important to orientate trainees to the andragogy-based safety training approach as they may be more accustomed to the passive pedagogical approach. This orientation aims to prepare workers to distinguish a proactive approach to learning from traditional reactive methods.

Safety learning in construction and engineering

In this section, we discuss common issues on safety learning in the construction and engineering industry and then showcase how the principles of andragogy can be used to address those issues.

Motivation to learning

The motivation to learn is influenced by external and internal factors. Extrinsic motivation is influenced by some externally-stimulated regulatory processes (Fazey & Fazey, 2001). For example, a worker learns about how to work safely at height because the learning process is enforced and regulated by the organisation. When the worker is motivated by a form of reward and punishment scheme implemented in the organisation, this is also considered as extrinsic motivation. In another case, although the worker may value the safety learning process, if the learning is initiated by the organisation, then this is still considered as extrinsic motivation.

Intrinsic motivation, on the other hand, is originated within the learner and is concerned with the task, that is, the learning subject itself. Intrinsic motivation is congruent with the learner's sense of self and is expressed in a desire to know, to achieve, and to be stimulated (Fazey & Fazey, 2001). Intrinsic motivation is essential for meaningful and worthwhile learning. It also promotes responsible and continuous learning (Garrison, 1997). Intrinsic motivation is crucial for safety learning. A study by Wilkins (2011) revealed that workers who undertook safety training of their own volition retain the knowledge acquired through the course far better than those who were paid for by their organisations or were compelled by regulations to do so (Wilkins, 2011).

At work, active participation, encouragement and guidance from the manager or the system applied in the organisation, that is, extrinsic motivational factors are certainly needed for any development effort to be productive. However, it needs to be kept in mind that no one can motivate a person towards self-development if the person has no desire to develop. Motivation must come from within, meaning that the desire to learn must be intrinsic. Peter Drucker (2008) said that people development is always self-development. For an organisation to assume full responsibility for the development of its employees is an idle boast. The responsibility rests with the people, their abilities, and their efforts. This motivation to self-develop will lead an individual to realise the need to develop others. It is in and through efforts to develop others that individuals eventually place demands on themselves and contribute to greater organisational performance. Interestingly, the best performers in any profession always look upon the individuals they have trained and developed as the proudest monument they can leave behind (Drucker, 2008). This is consistent with an assumption of andragogy which argues that internal motivators are key factors that motivate people to learn.

Supportive learning environment

It should be noted, however, that external motivators still play some role in generating this motivation to learn. In fact, extrinsic motivation may complement and enhance intrinsic motivation (Garrison, 1997). What construction organisations should do is to advance a climate and environment which values learning irrespective of pressure from an employer or certain regulations (Wilkins, 2011). For example, in conducting safety training the physical environment should support the learning process by making trainees feel at ease. A podium or a stage may make trainees feel that they are being talked down to. A room with rows of chairs may create an impression of suppression and passivity, while a room where trainees sit in small groups or around tables is considered to be more conducive for adult learners. More importantly, the psychological environment should cause trainees to feel accepted, respected, and supported. There should be freedom of expression and less status differentiation between the trainer and the trainee (Knowles, 1980). To encourage intrinsically motivated learning, learners must see opportunities to share control and to collaborate in the planning and implementation of the learning process (Garrison, 1997).

The behaviour of the trainer is also particularly important. Trainers convey in many ways whether they are interested in and respect their trainees or simply see the trainees as receivers of wisdom. Trainers who take the time to know their trainees, who are able to call them by name, and who demonstrate the act of listening attentively to what the trainees say are those who convey the attitude that promotes learning (Knowles, 1980).

This learning climate and environment can be extended beyond individual training programmes and applied to the entire organisation. It can be reflected in the organisation's interior, policies, procedures and leadership (Knowles, 1980). There are practical ways to help organisations create a climate that generates the motivation to learn. First, the attitude of top managers is paramount. Through what they say and how they behave, top managers establish norms that filter down through the organisation. Data collected from 202 companies in Spain found that CEOs who demonstrate transformational leadership promote organisations that are open to learning (Montes et al., 2005). As discussed in Chapter 4, transformational leaders demonstrate the following characteristics (Bass & Riggio, 2006):

- Idealised influence: Behaving in ways that allow them to serve as role models for their followers, thus making the leaders admired, respected and trusted.
- Inspirational motivation: Behaving in ways that motivate and inspire those around them by providing meaning and challenge to their followers' work.
- Intellectual stimulation: Stimulating their followers' efforts to be innovative and creative by questioning assumptions, reframing problems and approaching old situations in new ways.
- Individualised consideration: Giving special attention to each individual follower's needs for achievement and growth by acting as a coach or mentor.

Second, the hiring process should assess job candidates not only on the requirements of the job, but also how well they might fit into the values of the organisation. Third, it is necessary to have an effective socialisation process to help new employees learn the organisation's ways of doing things. Robbins et al. (2012) explained that socialisation can be accomplished through stories, rituals, material symbols and language used in the organisation. These socialisation methods are applicable in safety management as discussed below:

- Organisational stories which typically contain significant events or people that explain the organisation's heritage or celebrate people getting things done. These stories provide explanations and legitimacy for current practices, exemplifying what is important to the organisation and generating convincing pictures of an organisation's goals (Robbins et al., 2012). Consider the story of Gopalan Seenivasan published in Lend Lease's corporate website (Lend Lease, 2014). He was the Environmental, Health and Safety Manager in an international school project in Singapore. He instilled safety culture through his daily toolbox talks, mock-up safety drills and his prominent display of visual safety notices around the site. He provided first-class site amenities, using materials, plants and shrubs recycled from the demolition process. He also used daily exercise sessions to bring people together and promote a better quality of life. Lend Lease considered the site as a shining example of best practice across Lend Lease's Asia region, prompting other sites to implement similar initiatives.

- Rituals which are manifested in repetitive sequences of activities that express and reinforce the key values of the organisation, what goals are most important, and which people are important (Robbins et al., 2012). For example, Gammon Construction based in Hong Kong halted work across 110 worksites to hold an event called Stand Down so that everyone could focus on the issue of safety. The Stand Down was an opportunity for senior management to provide caring and visible leadership essential for instilling safety culture (Gammon Construction, 2013). A smaller scale ritual for safety is a tool box talk which can be done daily before work starts to remind workers about the importance of safety and about the key safety risks on that day.

- Material symbols which resemble an organisation's personality, such as formal, casual, fun and serious. These symbols can be in the form of office layouts, how employees dress, the elegance of furnishings, the class that top executives travel, executive perks and employee cafeterias. They generally convey the kinds of behaviour that are expected and appropriate (Robbins et al., 2012). An example of a material symbol to socialise safety is the regulation that enforces workers wearing basic personal protective equipment (PPE) like hard hat, safety shoes, and safety vest. Safety posters, warning signs, a safety induction room and good housekeeping are other examples of material symbols that promote safety.

- Language which is used within an organisation to describe equipment, key personnel, suppliers, customers or products related to their business. It is possible to take notice of the most commonly used words in the organisation to identify what is important and what people pay attention to (Robbins et al., 2012). Safety slogans are a simple, but potentially powerful tool to keep safety in the minds of workers. Examples of safety slogans are:
  - Injury and incident free.
  - Don't be a fool, use the proper tool.
  - Eyes are priceless, eye protection is cheap.
  - Know safety, no pain. No safety, know pain.
  - Safety glasses, all in favour say "EYE!"

Facilitating safety learning

Trainees' perceptions of their trainers' competence greatly influence the effectiveness of safety training programmes. On one hand, when the training course contents are being inadequately delivered, then the trainer should be retrained or replaced. On the other hand, when trainers are deemed to be competent, but are not trusted or respected by trainees, a more nuanced approach is needed. According to the principles of andragogy, rather than actively teaching in a more traditional, didactic way, the trainers should facilitate the trainees in the learning process (Wilkins, 2011).

Unlike children and teenagers, experienced workers have accumulated a great deal of experience. As such, trainers should use exercises that tie directly to work or life experience. The use of effective analogies can be a powerful technique for introducing new concepts and explaining complex procedures. Whenever feasible, training should also be tailored to specific audiences (Grupe & Connolly, 1995). Experienced workers also tend to be independent and self-directing, thus they are more interested in immediate, problem-centred approaches than in subject-centred ones. There are seven principles that can be used to teach this group of people (Kaufman, 2003):

1. Establish an effective learning climate where trainees feel safe and comfortable expressing themselves
2. Involve trainees in mutual planning of relevant methods and curricular content
3. Involve trainees in diagnosing their own needs to trigger internal motivation
4. Encourage trainees to formulate their own learning objectives, thus giving them more control over their learning
5. Encourage trainees to identify resources and devise strategies for using the resources to achieve their objectives
6. Support trainees in carrying out their learning plans
7. Involve trainees in evaluating their own learning through critical reflection

For example, in the USA, efforts have been made to develop safety training materials for Hispanic workers (Brown, 2003; Brunnette, 2004, 2005; Evia, 2011). Brunnette (2004) discovered that the use of a worker's participatory approach in the design, development and continuous evaluation stages are important because creative thinking can come from the workers who have hands-on experience. Brown (2003) suggested that rather than translating straight from English materials, using a native speaker who knows the topic well to write the Spanish text is recommended because this allows the writer to express things in a more culturally sensitive and less technical way. For example, in ergonomic training, Hispanic workers are more willing to say that they have discomforts (*molestias*) rather than pains (*dolores*). Another input is that the materials developed for workers in Spanish-speaking countries often used graphics that are somewhat humorous while still treating workers with respect, in contrast to English publications which tend to be extremely serious. Illustrated stories using a comic book format is also a popular form of reading material for Spanish-speaking people, particularly those who do not have advanced reading skills. Finally, it is important to pilot test drafts of materials with a subset of the group for which they are intended to gain ideas on how to make the materials more useful and appealing (Brown, 2003).

Burke et al. (2006) examined 147 safety training events which were attended by 20,991 participants and found that as the method of training becomes more engaging (going from passive methods such as lecture to experiential-based methods such as hands-on training that incorporate dialogue), the effect of training is greater for knowledge acquisition, safety performance improvements and the reduction of negative outcomes. The findings further indicated that the most engaging methods of safety training were, on average, approximately two times more effective than the moderately engaging methods and three times more effective than the least engaging methods with respect to workers' knowledge acquisition. The main point is that the highly engaging learning process which allows active participation of trainees and frequent interactions between trainers and trainees is superior to the passive learning process such as listening to lectures, watching videos and reading handbooks (Burke et al., 2006). This indicates the value of incorporating andragogy principles into safety training processes.

Furthermore, most workers have either heard a cautionary tale or personally experienced a work-related injury. Relating these experiences to the context of a training session can be useful to improve the effectiveness of training programmes (Wilkins, 2011). Consider the case of Donnie's accident in 2004 which occurred because of the neglect of safety procedures and the absence of PPE. An electrical explosion (called arc flash) caused Donnie to suffer from third degree burns to the muscle on both arms and hands and second degree burns to his face, head and neck. He was in a coma for more than a month, underwent a number of surgeries and was in therapy for one and a half years to learn using his hands and arms again. His story and videos have been used as a powerful message to showcase the importance of safety and the emotional impacts an accident could bring, not only to the victim, but also to people around him (Johnson, 2013).

Safety E-learning

A characteristic of adult learners is that they are typically protective of their time. As such, they are keen to pursue training only if it can be accommodated by their other priorities (Grupe & Connolly, 1995). The high-pressured and ever-changing work environment of the construction industry exacerbates the condition and causes an impression that safety training is delivered in an inconvenient way (Wilkins, 2011). As a result, in a traditional training setting, trainers have to manage time efficiently and maximise the time of the trainees in their learning. When there are a large number of trainees, additional assistants may be necessary, particularly when there are group or individual tasks involved. Furthermore, training sessions have to begin and end on time. Trainers should not penalise trainees who come on time by forcing them to wait for late attendees. Short discussions and overviews can be inserted after each topic to accommodate late comers (Grupe & Connolly, 1995).

E-learning is another way to address the time constraint issue of trainees (Wilkins, 2011). Today people realise the need to be educated and trained to respond to the demands imposed by their workplaces. At the same time, organisations realise that their human resource is a key source of competitive advantage in current business environments. This economic consideration creates a need to contain the cost of education and training. Due to the development of information technology and the growth of internet use, e-learning is seen to have an immense potential to address this challenge (Organisation for Economic Co-operation and Development, 2001). There are a number of advantages of e-learning. First, this mode of learning can be accessed from any location and at any time '24/7'. Second, e-learning can be uniquely adapted to learners with different learning styles, interests and cultural beliefs. Instead of having fixed content where everyone learns the same thing in a classroom environment, e-learning allows learners to independently access information, thus empowering learners in their learning process. Third, e-learning offers flexible pacing, which is ideal for training aimed at both new and experienced workers. Fourth, e-learning can be a cost-effective way of training the workforce (Acar et al., 2008; Loos & Diether, 2001; Organisation for Economic Co-operation and Development, 2001).

Currently the application of safety e-learning is still far from fulfilling its full potential. However, construction organisations should not dismiss this approach to learning as there are already promising achievements. Research has indicated that where learners have a choice, the online option is increasingly well accepted (Wagener & Zou, 2009). Consider the computer-based training module developed by BuildIQ and Virginia Tech Center for Innovation in Construction Safety and Health Research (Evia, 2011). With the input from Hispanic construction workers, who are also their target trainees, they developed a short training module on scaffold safety using a stop-motion animated video in Spanish language with a clear plot, practical recommendations and some humour. During the evaluation of the e-learning module, all workers who watched the video stated that the video is interesting and easy to understand. They were able to summarise the module and remembered the characters' names, indicating the enhanced memorability of the training module (Evia, 2011).

A study by Chung et al. (2005) found that e-learning is a practical method to study an undergraduate construction technology course. E-learning is also useful to help students improve their learning independence, learning efficiency and learning effectiveness. In this study, part-time students had better perceptions towards e-learning than full-time students due to differences in learning characteristics and requirements. However, most students prefer a combination of e-learning and face-to-face learning rather than a single mode of learning. They favoured the application of e-learning as a complement to the traditional face-to-face method of teaching and learning.

A study by McMahan et al. (2008) attempted to reduce haul truck accidents by improving worker training, using a virtual learning environment. The training consists of three phases: virtual tour, virtual inspection and simulation. The virtual tour introduces information necessary to conduct an inspection by guiding the workers around a haul truck, identifying parts to be inspected, and explaining defects to look for. The virtual inspection assesses retention as the worker navigates around a haul truck and identifies defects. During the simulation, the workers are shown a simulation which animates severe consequences of any overlooked defects to emphasise the importance of regular inspections. Feedback shows that virtual environment-based training has a potential to be more effective than traditional training methods, such as PowerPoint presentations and instructional videos. It is engaging and may even replace normal classroom training in the future (McMahan et al., 2008).

A more advanced development is the game technology-based safety training to provide trainees with hands-on experience in a virtual environment (Guo et al., 2012). Using this training platform, training modules for demonstrating the safe operation of a tower crane, mobile crane and pile driver have been developed. Evaluation shows that the training platform helps trainees understand plant operations and identify wrong operations in advance. It also improves operatives' ability to collaborate with each other and identify safety problems. This results in strengthening the relevant skills needed for operating the plant.

Nowadays social networking is considered as another platform for learning, particularly social learning. The idea of learning through communication and collaboration with other people is not new. As the technology evolves, millions of users are now connecting and collaborating with others on a variety of social networking websites such as LinkedIn, Facebook, and Twitter. A good example is Moodle, a free course management system to improve the interactive e-learning experience. Due to its open source licence, anyone can develop additional functionality and offer the new solutions back to the international Moodle community. Today many schools, organisations, and businesses around the world use Moodle to meet their online learning and social networking needs (Martinez & Jagannathan, 2010). Others use Facebook as their social networking and learning platform because it allows concurrent input of different agendas, approaches and priorities. Facebook allows users to create their own content, but it is also especially useful to quickly and easily disseminate information that was produced elsewhere. This ability to share information with a large number of people makes collaboration across time and space easy (Freishtat & Sandlin, 2010).

Nevertheless, we need to be aware that relying too heavily on a single approach, in this case e-learning, may be ineffective. A lack of understanding on the part of the trainees may easily go unnoticed. Indiscipline may lead to taking shortcuts in the learning process, relying on search engines rather than personal understanding to solve problems. Cheating on examinations is also possible (Wilkins, 2011). There are also situations where face-to-face delivery is preferable to e-learning, such as (Wagener & Zou, 2009):

- Learners have learning or language difficulties
- Insufficient or limited internet capability
- Companies are unable to afford e-learning induction packages or do not have the capacity to develop an e-learning programme
- Regulating authorities do not consider e-learning as a suitable mode of learning

Informal safety learning

So far in this chapter, our discussions on learning have focused on formal learning and structured training. There is another paradigm which argues that learning occurs informally in practice, not in classrooms. We will contrast the two and discuss the implications in the context of safety learning in the construction industry.

Nowadays many organisations consider learning simply as an acquisition of knowledge which can be achieved through instruction and training in a class-room setting, and that knowledge is available somewhere and learners need to acquire and store it in proper compartments of their minds (Gherardi, 2006; Gherardi & Nicolini, 2000, 2002). From this perspective, learning is mainly focused on its outcomes and very much taken as a "given". This view sees learn-ing as being achieved by simply 'plucking an item from the tree of knowledge' (Tsoukas & Mylonopoulos, 2004). Many educational and training approaches adopt the philosophy that views learning as a product that can be simply added to the mind or readily stored and transmitted via some kind of electronic tech-nology (Hager, 2004). Consequently, much organisational learning literature and studies have focused on the codification, packaging and dissemination of knowledge throughout organisations and workplaces (Styhre, 2006).

In the context of safety learning, the following steps are normally imple-mented by construction organisations to train their employees. Initially, they train everyone to a minimum standard, for example, by using the Safety White Card Course, which has been mandated in Australia, requiring individuals to complete the course before they do any construction activities. Passing this course, therefore, can be considered as a common denominator and an indication of basic safety competence in the Australian construction industry. Achieving this minimum requirement is far from sufficient for construction organisations who aspire to uphold safety as one of their priorities. Conse-quently, they 'upgrade' their employees' safety knowledge by authorising safety induction and safety training programmes, which typically explain various hazards that may be encountered at work and methods to do tasks safely. They have a toolbox talk at the start of the day to make people aware of particular activities that will be performed in the project, along with the potential hazards, during that day. In addition, these organisations also have a supervision system to monitor and remind people to follow safety procedures.

All these training and learning methods are, of course, useful for developing safety knowledge and skills. However, it is important to remember that safety learning is a process that does not happen instantaneously. There is a danger that construction organisations try to accelerate the rate of learning. Putting people in a room for 5 days to learn about safety and expect them to be experts, who would implement everything that they have learnt, is irrational. There is also an issue concerning the ineffectiveness of existing safety training programmes. A study has shown that a classroom-like training setting only has short-term impacts on safety performance. After a short period of time, workers tend to forget what they have learnt and as a result safety performance returns to where it was before (Laukkanen, 1999).

For this reason, it is important to view safety as a set of practices constituted by competences that a person learns through engagement and participation in daily activities (Baarts, 2009). Although the current approach to safety learning may seem practical and straightforward, a countervailing view would be that learning does not comprise a technological device, but something that is situated in local practices where people collaborate and cooperate to solve daily issues (Styhre, 2006). From this perspective, safety should be considered as the final outcome of a collective construction process. A safe workplace, therefore, is the result of constant engineering of diverse elements, for example, skills, materials and interpersonal interactions, which are integral to the work practices of various project stakeholders. In other words, learning about safety involves taking part in the social world, that is, learning takes place among and through others (Gherardi, 2006; Gherardi & Nicolini, 2002).

If we consider the reality of construction projects, many problems do not neatly fit into predetermined categories. This forces construction practitioners to employ novel solutions and creative strategies to manage non routine situations (Wadick, 2006). Workers are often required to make important decisions in a dynamic work environment. This kind of work environment demands that they talk to each other about potential hazards while they are in action and employ their own adaptive accident prevention strategies without waiting for site management approval (Saurin et al., 2008). This demonstrates that safety learning mainly occurs via peer learning or collaboration between peers and fellow professional groups. Seeing, saying, showing, telling, reading, reflective thinking and learning-by-using are how individuals acquire new safety skills and knowledge. In a sense, knowledge is integrated and distributed in everyday activities, and so learning cannot take place if participation in those activities is not possible (Styhre, 2006). Notwithstanding efforts to provide formal training, workers perceive that they receive real training on the job by watching more experienced workers, trying things out, or context-specific instruction from more experienced members of the site community (Rooke & Clark, 2005).

Therefore, it is understandable why many construction practitioners, especially workers, do not consider safety regulations, training and research as something beneficial for them. They believe that many safety rules do not address their real safety concerns, but are merely an attempt by powerful bureaucrats to dominate and subjugate their subjects. As a result, they resist such instructions by doing as little as they can only to comply, a far cry from the 'best practice' that those regulations, training and research often try to achieve (Wadick, 2006).

Many safety training methods are designed and implemented on the assumption that knowledge and learning are primarily individual and mental processes (Gherardi & Nicolini, 2000). Although the importance and necessity of this approach is undeniable, construction organisations should recognise the alternative paradigm where safety knowledge and learning are seen as social and cultural phenomena developed through interactions of individuals with each other and with non-human artefacts while working on sites (Wadick, 2006). In other words, learning should not only be seen as a product, but also a process where the learner is part of the environment. This view of learning emphasises the context and the influence of cultural and social factors in the learning process (Hager, 2004).

Informal safety learning essentially recognises that producing successful interventions to improve safety performance requires an understanding of local knowledge embedded in and conditioned by local tradition. This local knowledge and practice often differ from those represented in institutional codes through formal safety training efforts. Using an example of mobile phone use, Pink et al. (2010, 2014) showed a potential conflict between institutional knowledge and local practice, and stated that as mobile phones are ubiquitous a construction site decided to establish a few mobile phone zones to regulate their use. This institutional code, however, is impractical and causes ambiguity in practice. A safety manager explained that if there is an incident on site, it is expected that workers immediately report the incident. In fact, workers were encouraged to phone a number printed on their helmets in cases of safety emergencies on site. Upon observation, mobile phones were used by workers for the detailed coordination of safe work practices. Mobile phones were also used by managers to locate and move workers around the site easily and for institutionally recognised safety practices. This case shows that workers and managers innovated to create a 'safe' mobile phone behaviour on site, something which was learnt in practice due to the ambiguity of mobile phone use regulation. Therefore, it is recommended that safety interventions should recognise local practices through which safe work is enacted. Instead of focusing on standardised and institutionalised safe work practices, construction organisations should recognise and embrace local safe work practices. These local practices should be visibly rewarded so that safe work practices are made available and disseminated through formal channels. Instead of blindly following standardised practices, managers should be aware of the informal practices and unique requirements of the site (Pink et al., 2010; Pink et al., 2014).

Experiential safety learning

Research has also begun to recognise the importance of learning by doing. There has been a notable increase in the use of experiential training methods and rich interactive simulations, particularly in fields such as healthcare, the military and aviation. Such approaches provide the needed link between theory and practice while offering a holistic approach to learning, where ideas and knowledge are derived from and tested out in the experiences of learners (Nyateka et al., 2014). The construction and engineering industry has been slow in adopting this approach to learning, but some effort has been made to remedy this. For example, a study by Nyateka et al. (2014) presented two innovative wearable simulation suits to develop simulation-based training programmes for younger workers. First, a whole body simulation suit, called the Third Age Suit, simulates aspects of ageing as a mechanism for raising awareness within young designers about older driver characteristics and requirements. By wearing the suit, those young designers are able to experience what it is like to be in the shoes of older drivers. Through this understanding, design decisions become more in line with customer needs. Second, there are wearable devices called LUSKInS (Loughborough University Sensory and Kinaesthetic Interactive Simulations) which target younger workers' attitudes towards occupational health in order to reduce problems in later life. LUSKInS simulate key occupational ill-health conditions most prevalent due to working on construction sites, including dermatitis, hand–arm vibration syndrome, musculoskeletal disorders, noise-induced hearing loss and respiratory syndrome, and their consequential impacts on the daily living of the sufferers. By using such wearable simulation devices in safety training, the wearer is able to directly experience the difficulties, limitations and discomforts faced by sufferers, thus encouraging attitudinal and behavioural changes to occupational health matters (Nyateka et al., 2014).

Techniques for evaluating safety training and learning

So far we have discussed different forms of safety training, as well as formal and informal safety learning or 'knowing in practice'. This section discusses methods of evaluating safety training/learning programmes which is important to ensure that they are relevant and effective. Kirkpatrick (1979) developed a four-part process to evaluate the effectiveness of training programmes, which we argue can be applied in the context of safety training. This process is widely accepted and consists of four parts: *reaction, learning, behaviour* and *results*. Together, these four parts measure the immediate and long-term effects of learners' knowledge and capability improvement, as well as attitude and behavioural changes as the outcomes of training (Kirkpatrick & Kirkpatrick, 2006). We use Kirkpatrick's model as a basis to develop models for evaluating safety training programmes in the construction and engineering context as discussed in the following sections.

Part 1 – reaction to and satisfaction with the safety training programme

This part is a kind of satisfaction survey to measure how trainees feel about the various aspects of a training programme, including the topic, trainer, training approach and so forth. The goal of training is to make trainees essentially more effective, so this is a simple, yet practical way to find out whether or not a training programme is actually useful, according to the perceptions of the trainees. Measuring reaction is also important to ensure that trainees are motivated and interested in learning. If they do not like the training programme, then it is likely that they will not make an effort to learn. Reactions are valuable to obtain comments and suggestions, which will be helpful to improve future programmes. General guidelines for evaluating reaction are: determine what needs to be measured; design a form that will quantify reaction; encourage written comments and suggestions; attain an immediate response rate of 100%; seek honest reactions; develop acceptable standards; measure the reactions against the standards and take appropriate action and communicate the reaction results as appropriate (Kirkpatrick, 1979; Kirkpatrick & Kirkpatrick, 2006). An example of evaluating reactions to a safety training programme is given in Table 5.2.

Part 2 – safety learning outcomes

It is important to remember that a favourable reaction to a programme does not assure significant or desirable learning outcomes. A trainer may show

Table 5.2   A sample questionnaire to measure reaction to a safety training programme

Item	Strongly disagree	Disagree	Neither agree nor disagree	Agree	Strongly agree
The training topic/content was relevant to my needs.					
The training methods were effective for my learning.					
The trainer kept the session alive and interesting.					
The trainer maintained a friendly and helpful manner.					
The trainer delivered all the points clearly.					
I am satisfied with the performance of the trainer.					
I am satisfied with the training session.					
What is the best thing about the training session?					
How can the training session be improved?					

enthusiasm and use a variety of methods to make the learning process well accepted by the trainees. However, there could be a situation where a careful analysis reveals that the content has little value, although the trainer has presented it really well. At the end of the day, it is the outcome of learning that matters. The second part, therefore, measures the knowledge acquired, skills developed or attitudes changed as a result of the training. This part is not concerned with on-the-job use of the learning content, but focuses on the amount of content understood and absorbed by the trainees. Guidelines for evaluating learning outcomes are: where practical, use a control group that does not receive the training to compare with the training group; evaluate knowledge, skills, or attitudes both before and after the training; attain a response rate of 100%; and use the results of the evaluation to take appropriate action (Kirkpatrick, 1979; Kirkpatrick & Kirkpatrick, 2006).

Classroom performance, such as demonstrations, individual performance of the skill being taught, discussions and case study, may be used to evaluate learning. For example, in a course that is teaching safe working at heights, each trainee can be asked to demonstrate how to wear a body harness. In a first-aid training course, trainees can be evaluated on how they perform a cardiopulmonary resuscitation (CPR). In a scaffolding safety course, the trainer can present a case study and ask the trainees to prepare a safe method for erecting scaffolds, based on the case scenario. When these methods are planned and implemented properly, the trainers should be able to measure fairly objectively the amount of learning that has taken place.

When principles and facts are taught rather than techniques, learning effectiveness can be measured using the 'pre-training and post-training tests' method. Following the guidelines above, this test should be given to all trainees prior to the training. Whenever possible, the test should also be given to a control group which is comparable to the training group. The pre-training test will give trainers some understanding of the group prior to the training; thus they can focus on items most frequently misunderstood. After the training is over, the same test or its equivalent should be given to the trainees and also to the control group. A comparison of pre-training test and post-training test scores can then be made and statistically analysed to reveal the effectiveness of the training.

Part 3 – safety behavioural change

There is a major difference between knowing principles and techniques and applying them on the job. Therefore, Part 3 is a measure of the extent to which trainees change their on-the-job behaviour as a result of training. Guidelines for evaluating behaviour are: use a control group if feasible; allow enough time for change in behaviour to take place; survey or interview one or more of the following group: trainees, their managers, their subordinates and peers who often observe the trainees' behaviour on the job; use an appropriate sampling technique if the number of trainees is too large; repeat the evaluation at appropriate times and consider the cost of evaluating behavioural changes versus the potential benefits.

In the case of construction safety, there are a variety of methods that can be used to assess behavioural changes. A self-administered questionnaire can be disseminated to relevant groups to assess behavioural changes before and after the training. On-site observation is another method to detect the amount of safe and unsafe behaviour and to observe whether trainees implement what they have learnt into practice. Interviews with workers and managers are also useful to get insight concerning specific behaviours relating to the training outcomes. Apart from the trainees, the behavioural changes of their peers may also be observed/evaluated to assess the levels of influence of the trainees' changed behaviour on their peers. Chapter 7 provides details for undertaking on-site behaviour observation as a method of ethnographical research.

Part 4 – long-term results in change of safety culture

Part 4 is about measurement of the long-term results that occur due to training. This may include increased job satisfaction and morale, improvement in productivity, increased profits, reduced employee turnover, better client satisfaction and changes of work culture. In relation to safety management, it may also include reduction in number of incidents, accidents and injuries and improvement of safety culture; the safety culture maturity framework discussed in Chapter 3 is useful to assess this improvement. It is difficult to measure long-term results of safety training. Organisations should determine performance indicators that are relevant for their circumstances. In Chapter 8, we propose the balanced scorecard as a method to gauge the long-term results of strategic safety management which involves the implementation of safety training programmes.

Evaluation becomes more complex, difficult and expensive as it progresses from Part 1 to Part 4. There is a tendency to jump directly to Part 4, but understanding all four parts is necessary to obtain a complete picture regarding the effectiveness of a training and learning process (Kirkpatrick, 1979; Kirkpatrick & Kirkpatrick, 2006). Table 5.3 provides a framework which was based on the Kirkpatrick's model for the context of safety training programmes in construction and engineering.

Case study

Using the evaluation framework presented in the previous section, we assess the effectiveness of a safety leadership skill training programme, provided by Master Builders Association (MBA) of the Australia Capital Territory (ACT) in Australia. Below is a summary of the training programme and its evaluation.

Table 5.3   Framework for evaluating safety training in construction and engineering

Kirkpatrick's model	Safety training evaluation indicators	Sample questions or measurement instrument
Reaction	Trainees' reaction and satisfaction on different aspects of a safety training programme	• Were the learning aims and objectives made clear to you? • To what extent were the training topics relevant to your needs? • Was the time allocated to each training topic sufficient to cover the contents? • To what level were you satisfied with the performance of the trainer? • To what level were you satisfied with the training method? • To what level were you satisfied with the training content? • Are there other methods (such as e-learning, blended learning, social media based learning, or on-site mock-ups) that you think should be used for safety training and learning? • What is the best feature of the training programme? • What areas of the training programme can be improved? How?
Learning	Safety learning outcomes including understanding, knowledge and skills gained by the trainees	• What knowledge do you think you have gained from the training? Give examples. • What skills do you think you have developed as a result of the training? Give examples. • What attitude or perspectives do you think you have changed/improved as a result of the training? Give examples. • To what extent has your thinking changed in relation to construction safety issues as a result of the training? Give examples. • What changes or improvements have you had in your mental model and mindset as a result of the training? Give examples. • In addition, tests related to the facts, concepts, principles and techniques may be given before, during, and after the training,

(continued overleaf)

Table 5.3   *(continued)*

Kirkpatrick's model	Safety training evaluation indicators	Sample questions or measurement instrument
Behaviour	Safety behavioural changes in trainees and their peers due to safety training	• Have you applied what you learnt in the training in your work practices? If you have, can you describe how you have applied it? And why? Give examples. • What are the barriers that hinder you from applying what you have learnt in the safety training? Give examples. • How has your safety behaviour changed as a result of the training? Give examples. • How have your peers' safety behaviour changed as a result of the influence from your changed safety behaviour? Give examples. • Question to ask peers, supervisors and subordinates: • What changes have you seen from the trainee's safety behaviour as a result of her/his training? Give examples. • What changes have you made in your own safety behaviour as a result of the influence of the trainee's changed safety behaviour? Give examples.
Result	Long-term results in changes of safety culture and other relevant long-term indicators	In order to measure the long-term changes of safety culture, the framework for measuring safety culture maturity given in Chapter 3 can be used before and after the training. The 'after training' measurement can be done at different time points, depending on the needs. The measurement includes three aspects – psychological, behavioural, and corporate. Note that the emphasis here is on the *long-term* changes that have happened as a result of the training. Other indicators that can be used for before-and-after training comparisons are: • enhanced safety climate (sample questions are given in Chapter 3, Table 3.1) • reduced incidence and injury rates and impacts • reduced lost time due to incidents and injuries • improved client satisfaction • increased job satisfaction and morale • improvement in productivity • increased profits • reduced employee turnover

Table 5.4   Description of the MBA training course: a trainee's perspective

The October 2013 training course was facilitated by XXX and YYY, and consisted of approximately 20 participants undertaking three half-day (12.30 pm – 4.00 pm) classroom sessions followed by a one-hour site visit to a commercial construction site in Canberra. The classroom activities included PowerPoint presentations to explain the rationale behind the training and the development of the concepts that form the basis of the programme. PowerPoint presentations were used by the facilitators to highlight pertinent concepts. Videos were used as case studies, and to understand and develop participants' observation skills. In addition, classroom and role playing activities were also undertaken to develop participants' skills. The site visit, that is, on-site coaching session, was used to demonstrate the observations and conversations skills developed during the training.

The training programme

The MBA of ACT's Safety Leadership Advanced Observations and Conversations Skills Training programme was developed and prepared by an independent consulting company specialising in construction safety training and provided to the ACT building and construction industry through Master Builders ACT Group Training. The training course was first held in April 2012 and has subsequently run for nine times. A total of 170 people who were working in the building and construction industry in the ACT and the greater Canberra region have attended and completed the training course. The course consists of three half-days in a classroom setting followed by a two-hour on-site coaching session. Overall, the course covers the following topics and contents:

- Introduction to psychology and culture of safety and risk through safety observations and conversations
- Advanced hazard and risk identification
- Culture and values

A descriptions of the training course, from a trainee perspective, is provided in Table 5.4.

Aims of the evaluation

Evaluation will help to determine the cost–benefit ratio of training programmes, provide feedback on the effectiveness of training and determine how much the trainees have benefited (Phillips, 2011). Based on the requirements set by the Master Builders of the ACT, the aims of this evaluation are to:

- Evaluate the extent that the training programme is working, or otherwise, from trainee perspectives;

- Assess if the training programme has had an impact on changing the participants' behaviour and the organisation's safety culture;
- Identify areas for future improvements to the training programme;
- Develop strategies for future improvement.

Methodology and processes of the evaluation

Two methods were used to evaluate the training programme: quantitative survey and qualitative interviews. The survey questionnaire, as shown in Table 5.5 was to gain the trainees' reactions to the training programme. This survey was implemented at the end of each interview.

To obtain results for the other three parts, that is, learning outcomes, behavioural changes and long-term results, semi-structured one-to-one face-to-face interviews with trainees were conducted. A qualitative approach is appropriate when a detailed understanding of a process or experience is required. The research process that has been followed includes the following steps:

- Conduct the interviews and write down the interviewees' answers.
- Audio record the interviews.
- Obtain a transcript of the interviews.
- Code the interview data using NVivo 10 computer software.
- Analyse the interview data using themes based on the research questions.
- Prepare and distribute (at the end of the interviews) the survey questionnaires.
- Analyse the survey questionnaire responses.
- Interpret and report the meanings and outcomes of the data analysis.

Table 5.5   Survey questionnaire for reaction to training programme

No.	Question	Strongly disagree	Disagree	Neither agree nor disagree	Agree	Strongly agree
1	The training topic was relevant to your needs.					
2	The training objectives were made clear to you.					
3	The training methods were effective for your learning.					
4	The trainer kept the session alive and interesting.					
5	The trainer maintained a friendly and helpful manner.					
6	The trainer delivered all the points clearly.					
7	The trainer met the objectives of the training.					
8	You are satisfied with the performance of the trainer.					
9	You are satisfied with the overall training session.					
10	What is the best feature of this training?					
11	What can be improved in the training programme?					

While the overall evaluation question is: "Does it work?", the following questions were put to the interviewees:

1. What knowledge have you learnt (or skills have you developed) as a result of the training?
2. Do you believe your attendance and participation in the course has resulted in a 'culture change' within your organisation, your project team and to yourself?
3. Have you changed any safety behaviour as a result of the training?
4. Can you provide examples of how the behaviour has changed and why it has changed?
5. What suggestions do you have for future improvement of the training programme?
6. Do you have any other comments or observations?

The evaluation results

Twenty-six interviews and surveys were undertaken and the key findings are:

- 96% of participants had a positive reaction to the course.
- 77% of participants gained and retained new safety knowledge.
- 88% of participants acquired new skills.
- 35% of participants believed they had changed their safety behaviours.
- 88% of participants improved their safety behaviours.
- 62% of participants believed their new safety behaviours had led to improved safety behaviours of their fellow workers.
- 50% of participants identified a change in the safety culture of their project team.
- 73% of participants believed their attendance at the course led to an improved safety culture within their organisation.

Suggestions for improving the training programme as provided by the interviewees included:

- More time for on-site coaching session as part of the training.
- More effective use of the on-site time.
- More intervals of post-interview evaluations.
- Inclusion of a post-training refresher course.
- Introducing the MBA of the ACT training course to subcontractors to further improve the whole project team safety culture.
- Run a similar training course designed for other stakeholders within the construction industry such as clients, consultants and unions, in order to ensure the whole construction industry is on the same page.
- In future, evaluation research should include pre-training and post-training surveys and/or interviews and ongoing monitoring and feedback, to provide more comparative and valuable data and lead to further course improvements.

It was also recommended that the MBA of the ACT's training course be continued and the course improvement suggestions made in the evaluation report be considered in the training programme design. More comprehensive evaluation procedures should also be implemented concurrently with the training course by using pre-test–post-test comparative research methods.

Conclusions

In any organisation the most important resource is the workforce. Today human resources have become the source of competitive advantage for organisations across industries. Developing their human resource, therefore, is one of the key management strategies in contemporary organisations. Training and learning are seen as the best ways to develop human resources. Two learning approaches have been discussed in this chapter: pedagogy and andragogy. Although pedagogy is necessary, due to the characteristics of construction personnel and the nature of construction safety training/learning, the principles of andragogy should also be used as they have been proved to be more effective than the pedagogy approach alone. We have explained how andragogy can be used in construction safety training and learning.

This chapter has also discussed the motivation to learn as well as safety e-learning and its potential to reduce costs and to empower learners in their learning process, thus allowing them to internalise their learning and develop their intrinsic motivation to learn. Studies have shown that intrinsically motivated learners learn more effectively than those motivated by external factors.

This chapter has also discussed formal learning and informal learning, in relation to construction safety, in the classroom and in the workplace. Organisations should recognise the reality that learning occurs in practice through observing, listening and interacting with people and artefacts at work. No matter how good the formal training programme is, when the reality in the workplace does not reflect what is taught during training, learning will be stalled. Therefore, safety learning and safety culture development should go hand in hand.

Lastly, a four-part safety training evaluation framework and process, consisting of reaction to training programmes, learning outcomes, behavioural changes and long-term results, has been recommended for evaluating safety training programmes. Part 1 'Reaction' measures learners' satisfaction in different aspects of the safety training programme. Part 2 'Learning' measures the safety knowledge acquired, safety skills improved, or safety attitudes changed as a result of the training. Part 3 'Behavioural Change' measures the extent to which trainees change their on-the-job safety behaviour because of training. The trainee's peers' safety behavioural change, as a result of the influence of the trainee's changed behaviour, may also be measured. Part 4 'Results' measures the effects of the training programme on long-term results of the organisation, such as a more mature safety culture. The application of this four-part evaluation formwork is demonstrated by using a real case example.

Generally speaking, learning starts from birth and never ends, and the same is applicable for safety learning in that workers should continuously acquire new knowledge and skills in the classroom, in the workplaces, in social interaction and in self-reflection. Only with such consistent effort can safety performance be continuously improved.

References

Acar, E., Wall, J., McNamee, F., Carney, M., & Öney-Yazici, E. (2008). Innovative safety management training through e-learning. *Architectural Engineering and Design Management, 4*(3–4), 239–250.

Albert, A., & Hallowell, M. R. (2013). Revamping occupational safety and health training: Integrating andragogical principles for the adult learner. *Australasian Journal of Construction Economics and Building, 13*(3), 128–140.

Argote, L. (1999). *Organizational Learning: Creating, Retaining and Transferring Knowledge*. Norwell, MA: Kluwer Academic.

Baarts, C. (2009). Collective individualism: The informal and emergent dynamics of practising safety in a high-risk work environment. *Construction Management and Economics, 27*(10), 949–957.

Bass, B. M., & Riggio, R. E. (2006). *Transformational Leadership* (2nd ed.). Mahwah, NJ: Lawrence Erlbaum Associates.

Brown, M. P. (2003). An examination of occupational safety and health materials currently available in Spanish for workers as of 1999. In The National Academies (Ed.), *Safety is Seguridad* (pp. 83–92). Washington, DC: The National Academies Press.

Brunnette, M. J. (2004). Construction safety research in the United States: Targeting the Hispanic workforce. *Injury Prevention, 10*(4), 244–248.

Brunnette, M. J. (2005). Development of educational and training materials on safety and health: Targeting Hispanic workers in the construction industry. *Family & Community Health, 28*(3), 253–266.

Burke, M. J., Sarpy, S. A., Smith-Crowe, K., Chan-Serafin, S., Salvador, R. O., & Islam, G. (2006). Relative effectiveness of worker safety and health training methods. *American Journal of Public Health, 96*(2), 315–324.

Chung, J. K. H., Shen, G. Q. P., Leung, B. Y. P., Hao, J. J. L., Hills, M. J., Fox, P. W., & Zou, P. X. W. (2005). Using e-learning to deliver construction technology for undergraduate students: A case study in Hong Kong. *Architectural Engineering and Design Management, 1*(4), 295–308.

Drucker, P. (2008). *Management* (Revised ed.). Pymble, NSW, Australia: HarperCollins.

Evia, C. (2011). Localizing and designing computer-based safety training solutions for Hispanic construction workers. *Journal of Construction Engineering and Management, 137*(6), 452–459.

Fazey, D. M. A., & Fazey, J. A. (2001). The potential for autonomy in learning: Perceptions of competence, motivation and locus of control in first-year undergraduate students. *Studies in Higher Education, 26*(3), 345–361.

Freishtat, R. L., & Sandlin, J. A. (2010). Facebook as public pedagogy: A critical examination of learning, community, and consumption. In T. T. Kidd & J. Keengwe (Eds.), *Adult Learning in the Digital Age: Perspectives on Online Technologies and Outcomes* (pp. 148–162). Hershey: Information Science Reference.

Gammon Construction. (2013). *The Power of 10, Sustainability Report 2012*. Hong Kong: Gammon Construction.

Garrison, D. R. (1997). Self-directed learning: Toward a comprehensive model. *Adult Education Quarterly, 48*(1), 18–33.

Gherardi, S. (2006). *Organizational Knowledge: The Texture of Workplace Learning*. Malden, MA: Blackwell Publishing.

Gherardi, S., & Nicolini, D. (2000). To transfer is to transform: The circulation of safety knowledge. *Organization, 7*(2), 329–348.

Gherardi, S., & Nicolini, D. (2002). Learning the trade: A culture of safety in practice. *Organization, 9*(2), 191–223.

Grupe, F. H., & Connolly, F. W. (1995). Grownups are different: Computer training for adult learners. *Journal of Systems Management, 46*(1), 58–64.

Guo, H., Li, H., Chan, G., & Skitmore, M. (2012). Using game technologies to improve safety performance of construction plant operations. *Accident Analysis & Prevention, 48*, 204–213.

Hager, P. (2004). Conceptions of learning and understanding learning at work. *Studies in Continuing Education, 26*(1), 3–17.

Johnson, D. (2013). Donnie's Accident Retrieved 21 February, 2014, from http://www.donniesaccident.com/

Joo, B.-K. (2005). Executive coaching: A conceptual framework from an integrative review of practice and research. *Human Resource Development Review, 4*(4), 462–488.

Kalliath, T., Brough, P., O'Driscoll, M., Manimala, M. J., & Siu, O.-l. (2010). *Organisational Behaviour: A Psychological Perspective for the Asia Pacific*. North Ryde, NSW: McGraw-Hill Australia.

Kaufman, D. M. (2003). ABC of learning and teaching in medicine: Applying educational theory in practice. *British Medical Journal, 326*, 213–217.

Kirkpatrick, D. L. (1979). Techniques for evaluating training programs. *Training and Development Journal*, June, 178–192.

Kirkpatrick, D. L., & Kirkpatrick, J. D. (2006). *Evaluating Training Programs: The Four Levels* (3rd ed.). San Francisco, CA: Berrett-Koehler.

Knowles, M. S. (1980). *The Modern Practice of Adult Education: From Pedagogy to Andragogy*. New York: Cambridge.

Knowles, M. S., Holton III, E. F., & Swanson, R. A. (1998). *The Adult Learner: The Definitive Classic in Adult Education and Human Resource Development* (5th ed.). Woburn, MA: Butterworth-Heinemann.

Laukkanen, T. (1999). Construction work and education: Occupational health and safety reviewed. *Construction Management and Economics, 17*(1), 53–62.

Lend Lease. (2014). Safety Construction Leader: Gopalan Seenivasan Retrieved 20 February, 2014, from http://www.lendlease.com/worldwide/sustainability/~/media/91BD53FD54DB43A5BE5536583D394B20.ashx

Loos, G., & Diether, J. W. (2001). Occupational safety and health training on the Internet. *AAOHN Journal, 49*(5), 231–234.

Martinez, M., & Jagannathan, S. (2010). Social networking, adult learning success and Moodle. In T. T. Kidd & J. Keengwe (Eds.), *Adult Learning in the Digital Age: Perspectives on Online Technologies and Outcomes* (pp. 68–80). Hershey: Information Science Reference.

McMahan, R. P., Bowman, D. A., Schafrik, S., & Karmis, M. (2008). Virtual environment training for preshift inspections of haul trucks to improve mining safety. Paper presented at the First International Future Mining Conference, Sydney.

Montes, J. L., Moreno, A. R., & Morales, V. G. (2005). Influence of support leadership and teamwork cohesion on organizational learning, innovation and performance: an empirical examination. *Technovation, 25*(10), 1159–1172.

Nyateka, N., Gibb, A., Dainty, A. R. J., Bust, P. D., & Cook, S. (2014). The role of simulation-based learning in the occupational health training of younger construction workers Paper presented at the CIB W099 Achieving Sustainable Construction Health and Safety, Lund, Sweden.

Organisation for Economic Co-operation and Development. (2001). *E-Learning: The Partnership Challenge.* Paris: OECD.

Phillips, J. J. (2011). *Handbook of Training Evaluation and Measurement Methods* (3rd ed.). Abingdon, UK: Routledge.

Pink, S., Morgan, J., & Dainty, A. R. J. (2014). Safety in movement: Mobile workers, mobile media. *Mobile Media & Communication, 2*(3), 336–351.

Pink, S., Tutt, D., Dainty, A. R. J., & Gibb, A. (2010). Ethnographic methodologies for construction research: Knowing, practice and interventions. *Building Research & Information, 38*(6), 647–659.

Robbins, S. P., Bergman, R., Stagg, I., & Coulter, M. (2012). *Management* (6th ed.). Australia: Pearson.

Rooke, J., & Clark, L. (2005). Learning, knowledge and authority on site: A case study of safety practice. *Building Research & Information, 33*(6), 561–570.

Ruona, W. E. A., & Gibson, S. K. (2004). The making of twenty-first-century HR: An analysis of the convergence of HRM, HRD, and OD. *Human Resource Management, 43*(1), 49–66.

Saurin, T. A., Formoso, C. T., & Cambraia, F. B. (2008). An analysis of construction safety best practices from a cognitive systems engineering perspective. *Safety Science, 46*(8), 1169–1183.

Styhre, A. (2006). Peer learning in construction work: Virtuality and time in workplace learning. *Journal of Workplace Learning, 18*(2), 93–105.

Tsoukas, H., & Mylonopoulos, N. (2004). Introduction: Knowledge construction and creation in organizations. *British Journal of Management, 15*(S1), S1–S8.

Wadick, P. (2006). Learning safety in the building industry. Retrieved from http://www.cfmeu-construction-nsw.com.au/pdf/pwreslearnsafetybldgind.pdf

Wagener, E., & Zou, P. X. W. (2009). E-learning for construction safety training in the Australian construction industry Paper presented at the CIB W099 Conference on Working Together: Planning Designing and Building a Healthy and Safe Construction Industry, Melbourne, Australia.

Wilkins, J. R. (2011). Construction workers' perceptions of health and safety training programmes. *Construction Management and Economics, 29*(10), 1017–1026.

6 Safety in Design, Risk Management and BIM

This chapter discusses why and how safety should be considered during design stages (including architectural design and engineering design), from different perspectives such as legal, engineering, risk management and information technology. Conventional approaches to safety management have focused on methods, tools, procedures, unsafe acts, on-site work conditions and hazards; these focuses have been mainly at the construction stage. Although these approaches are needed, researchers, governments and practitioners have acknowledged the importance of examining, minimising and mitigating safety risks in the design stage. As commonly understood, architectural and engineering designs have significant impacts on the construction, operation and maintenance of facilities and infrastructure (including buildings, bridges, roads, highways, etc.). Research and practice have shown that many safety risks can be eliminated, mitigated, or reduced at the design stage if proper analysis and action are carried out during the design process.

This chapter focuses on the theory and practice of safety in design, including the definition, importance, policies and legal requirements, processes and techniques and success factors. Risk management is used as a method for safety in design. Case studies are used to demonstrate how safety in design can be applied in practice. The chapter also explains the use of building information modelling (BIM) as a tool to identify safety risks in the design stage.

What is safety in design?

Safety in design is the integration of hazard identification and risk assessment methods early in the design process to eliminate or minimise the risks of injury

Strategic Safety Management in Construction and Engineering, First Edition.
Patrick X.W. Zou and Riza Yosia Sunindijo.
© 2015 John Wiley & Sons, Ltd. Published 2015 by John Wiley & Sons, Ltd.

throughout the life of the building or structure being designed, including construction, use, maintenance and demolition (Consultants' Health and Safety Forum, 2012). In the USA, this concept is called the *Prevention through Design* (PtD), which is defined as the practice of anticipating and 'designing out' potential occupational safety and health hazards and risks associated with new processes, structures, equipment or tools, and organising work, taking into consideration the construction, maintenance, decommissioning and disposal/recycling of waste material, and recognising the business and social benefits of so doing (Schulte et al., 2008). In Australia, instead of using the term 'safety in design', an alternative term of 'safe design' and a broader definition was given by Safe Work Australia (2012c) in its Safe Design of Structures Code of Practice. The Code states that 'safe design means the integration of control measures early in the design process to eliminate or, if this is not reasonably practicable, minimise risks to health and safety throughout the life of the structure being designed. The safe design of a structure will always be part of a wider set of design objectives, including practicability, aesthetics, cost and functionality. These sometimes competing objectives need to be balanced in a manner that does not compromise the health and safety of those who work on or use the structure over its life. Safe design begins at the concept development phase of a structure when making decisions about: the design and its intended purpose; materials to be used; possible methods of construction, maintenance, operation, demolition or dismantling and disposal; and what legislation, codes of practice and standards need to be considered and complied with' (Safe Work Australia, 2012c, p. 6).

As indicated by the definitions given above, there are two key principles of safety in design. First, the concept aims to identify and eliminate or mitigate safety risks early in the design stage. Eliminating hazards at the design stage is often easier and cheaper to achieve than making changes later, when the hazards become real risks in the workplace (Safe Work Australia, 2012c). The concept is based on the popular premise in project management as shown in Figure 6.1, which argues that it is easier and cheaper to influence a project early in the project lifecycle rather than in later stages. Second, the concept encourages relevant stakeholders to identify potential safety risks not only in the construction stage, but also during the operation and maintenance of the facility or structure and during the dismantling process at the end of the lifetime of the facility or structure.

Schulte et al. (2008) provide a brief description of the historical development of the safety-in-design concept, particularly in the US context. The link between design and safety has actually been recognised since the beginning of 1800s, involving inherently safer design and the widespread implementation of guards for machinery, controls for elevators and boiler safety practices. Following this, there was enhanced design for ventilation, enclosures, system monitors, lockout controls and hearing protectors. More recently, there has been the development of chemical process safety, ergonomically engineered tools, chairs, work stations, lifting devices, and many others. Since the 1970s,

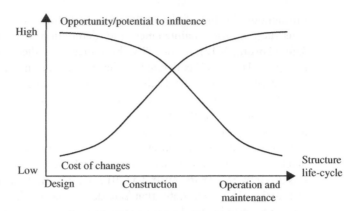

Figure 6.1   Cost of changes and opportunity to influence throughout project life cycle

the safety in design concept has been manifested in various management efforts. One of those efforts was the Safety and Health Awareness for Preventive Engineering (SHAPE), a collaborative programme between the US National Institute for Occupational Safety and Health (NIOSH), engineering professional societies and engineering schools to enhance the education of engineering students in health and safety, in which a series of nine instructional modules has been produced (Centers for Disease Control and Prevention, 2013). The Accreditation Board for Engineering and Technology also voiced the need to adopt new evaluation criteria, calling for health and safety curricular objectives and specific requirements in design, laboratory and professional practice instruction. The US Occupational Safety and Health Administration (OSHA) established the Alliance Program Construction Roundtable to bring construction-related alliance programme participants together to discuss and share information on workplace safety and health (WSH), in which one of the topics is about designing for construction safety. Other efforts include the issuance of the Process Safety Management of Highly Hazardous Chemical Standards, the National Safety Council's Integrating Safety through Design Symposium, the establishment of the Institute for Safety through Design and the Whole Building Design approach (Schulte et al., 2008).

In the context of infrastructure and building projects, safety in design should always be considered as part of a wider set of design objectives, including practicability, aesthetics, cost and functionality. Essentially, these sometimes competing objectives need to be balanced in a manner that does not compromise the health and safety of those who work on or use the facility or structure during its life. Designers, in this case, include architects and other designers who contribute to, or have overall responsibility for, any part of the design; engineers; contractors who carry out design work as part of their contribution to a project; temporary work engineers, including those designing formwork, falsework, scaffolding and sheet piling; and others who specify how alteration, demolition and dismantling work is to be carried out (Safe Work Australia, 2012c).

Why is it necessary to implement safety in design?

There are several reasons for implementing safety in design, including legal requirements, cost–benefits gains, and cause–effect logic. Traditionally, there is a notion that contractors are the main parties responsible for establishing safe construction processes and for enforcing safety on site, However, we need to realise that designers also play an important role in influencing construction worker safety. As shown in Figure 6.1, design has significant impact on construction methods and processes and the use and maintenance of the physical facilities. Safety should be considered early in the project lifecycle because the ability to influence safety is progressively lost as the project moves into the construction stage. Design also dictates how a project will appear and, consequently, how building components will be assembled (Gambatese & Hinze, 1999). The UK International Labour Office (ILO) recommended that designers should exercise care by excluding anything in the design which would cause safety hazards during the construction and maintenance stages of a facility (International Labour Office, 1992).

Research on causes of construction accidents have revealed that many of those undesired events originated from upstream or in the earlier stages of the project lifecycle. One of those studies states that two-third (2/3) of fatal accidents on construction sites are due to shortcomings in design (European Foundation, 1991). A study in the USA found that 42% of fatalities and 22% of disabling injuries were linked to design (Behm, 2005). Hale et al. (2007) estimated that 20–60% of accidents have at least one root cause which can be attributed to design error. In addition, a proof of safe design is increasingly becoming an important criterion to operate in certain markets. Contracts and liability claims increasingly focus on the ability of an organisation to prevent damage and injury. Ethical considerations and concern for reputation are other factors that compel organisations to embrace safety in the design concept.

Safe Work Australia (2012c) highlights that eliminating hazards at the design or planning stage is often easier and cheaper to achieve than making changes later when the hazards become real risks in the workplace. Furthermore, safe design can result in many benefits, including more effective prevention of injuries and illnesses; improved useability of structures; improved productivity and reduced costs; better prediction and management of production and operational costs over the lifecycle of a structure; innovation, in that safe design can demand new thinking to resolve hazards that may occur in the construction stage and in end use of building and structure.

Zou et al. (2009) summarised the following reasons as to why we need to consider safety in design:

- It is a requirement mandated by laws and regulations in many countries. While more details are given in the following section, some examples are provided here. The UK's Construction (Design and Management) (CDM) Regulations requires designers and clients in the UK construction industry to eliminate hazards in the design stage in order to make buildings safer to construct, use, maintain and demolish. The American Society of Civil Engineers (ASCE) states that engineers shall have responsibility for recognising that safety and constructability are important considerations when preparing construction plans and specifications. The Australian government also places similar responsibilities on designers through its Work Health and Safety Act 2012.

- In construction project management, many risks could be eliminated (and opportunities created) if proper analysis is carried out at the design stage. According to the Australian National Occupational Health and Safety Commission (2003), 42% of the 210 identified workplace deaths definitely or probably had design related issues involved. As such, designers should carry out comprehensive investigation of site conditions, articulate the clients' needs in a technically competent way and within the limitation of the clients' resource, work collaboratively to develop sound programme schedule and cost planning, and minimise defective designs.
- Identifying and eliminating risks at the design stage is a key to effective cost and managerial control (Andres, 2002), which may lead to many benefits, such as improved productivity, avoiding expensive retrofitting to correct design shortcomings, significant reduction in environmental damage, reducing costs and improving the usability of the facility.

The next section discusses safety in design from policy and guideline perspectives to further highlight the importance of considering safety in the design stage.

Safety in design policies and guidelines

Several countries have developed and implemented policies and guidelines on safety in design and four of such countries are chosen for discussion: the UK, Australia, the USA and Singapore.

In the UK

In the UK, the government has recognised the importance of safety in design by introducing the CDM Regulations 1994, which came into force in 1995. This has since been replaced by CDM Regulations 2007 (CDM2007), which also bring the Construction (Health Safety and Welfare) Regulations 1996 into a single set of regulations (CDM Regulations, 2007). The CDM Regulations are about focusing attention on effective planning and management of construction projects from the design stage onwards. Its aim is for health and safety considerations to be treated as a normal part of a project's development, not an afterthought. The objective of the CDM Regulations is to reduce the risk of harm to those who build, use, maintain and demolish structures (Health and Safety Executive, 2014).

Section 11 of the CDM2007 states the duties of designers in maintaining and improving health and safety in the construction industry. It requires all project designers, *so far as is reasonably practicable*, to eliminate hazards and risks during design and provide information about remaining risks. CDM2007 defines a designer as anyone who prepares designs for construction work, including variations. The designer includes anyone who arranges for their employees or other persons under their control to prepare designs.

Part 4 of the CDM2007 also spells out in detail each party's duty and responsibility in relation to safety during the design stage, including safe places of work, good order and site security, stability of structures, demolition or dismantling, explosives, excavations, fire, lighting, fresh air, temperature, weather protection and emergency procedures. In principle, when hazards cannot be removed by designers, design solutions should reduce the overall risk to an acceptable level (CDM Regulations, 2007). In practice, this means that when a potential hazard in the design is identified and cannot be removed, designers must reduce the likelihood of a harmful occurrence, reduce the potential severity of harm resulting from an occurrence, reduce the number of people exposed to the harm and reduce the exposure to harm in terms of duration or frequency (Howarth & Watson, 2009). CDM2007 also requires designers to provide information to assist other stakeholders to identify and manage any significant remaining risks that have not been designed out.

In Australia

The Australian government identifies health and safety by design as one of the national action areas in the Australian Work Health and Safety Strategy 2012–2022. The strategy recognises that the most effective and durable means of creating a healthy and safe work environment is to eliminate hazards and risks during the design of new plants, structures, substances and technology, and of jobs, processes and systems. This design process needs to take into account hazards and risks that may be present at all stages of the lifecycle of structures, plants, products and substances. The strategic outcomes of this action area are that (1) structures, plants and substances are designed to eliminate or minimise hazards and risks before they are introduced into the workplace and (2) work, work processes and systems of work are designed and managed to eliminate or minimise hazards and risks (Safe Work Australia, 2012a). Furthermore, the Work Health and Safety Act (WHS Act) 2011 section 22 states that designers have a responsibility to ensure, *so far as is reasonably practicable*, that their products are without risks to health and safety when used at a workplace throughout their entire lifecycle. The WHS Act further emphasises that designers must carry out tests and examinations sufficient to ensure that their products meet work health and safety requirements. Adequate information must also be provided to those for whom the product was designed, about its intended purpose, test results, and any conditions necessary to ensure that it is safe and without risk to health or safety, when used for its intended purpose (Safe Work Australia, 2012b).

In addition, the Safe Work Australia's Code of Practice (2012c) in relation to the safe design of structures specifies the duties of everyone involved in the design process. A designer is a person conducting a business of undertaking (PCBU), whose profession, trade or business involves him/her in preparing sketches, plans or drawings for a structure, including variations to a plan or changes to a structure and making decisions for incorporation into a design that may affect the health or safety of persons who construct, use or carry out other activities in relation to the structure. Designers include:

- Architects, building designers, engineers, building surveyors, interior designers, landscape architects, town planners and all other design practitioners contributing to, or having overall responsibility for, any part of the design (e.g. drainage engineers designing the drain for a new development)
- Building service designers, engineering firms or others designing services that are part of the structure such as ventilation, electrical systems and permanent fire extinguisher installations
- Contractors carrying out design work as part of their contribution to a project (e.g. an engineering contractor providing design, procurement and construction management services)
- Temporary works engineers, including those designing formwork, false-work, scaffolding and sheet piling
- Persons who specify how structural alteration, demolition or dismantling work is to be carried out.

Clients also have specific duties under the WHS Regulations, to consult with the designer, so far as is reasonably practicable, about how to ensure that health and safety risks arising from the design during construction are eliminated or minimised and provide the designer with any information that the client has in relation to the hazards and risks at the site where the construction work is to be carried out.

In the US

In the US, currently there is no regulation on safety in design, although initiatives and standards exist (NIOSH, 2014). The US NIOSH (NIOSH, 2014) is leading a national PtD initiative to promote the concept and highlight its importance in all business decisions. The national PtD initiative requires input from key industries including agriculture, forestry and fishing; mining; construction; manufacturing; wholesale and retail trade; transportation, warehousing and utilities; services and healthcare and social assistance. The PtD initiative is framed within four functional areas: research, practice, education and policy. Research focuses on questioning current practices to generate improvement. Practice focuses on encouraging businesses to demand safer designs and motivating design professionals to increase their awareness of those design features that can impact worker health and safety. Education focuses on promoting PtD through the augmentation of curricula and by stimulating professional accreditation programmes to value PtD issues and to include them in competency assessments. Policy focuses on supporting the other functional areas and providing incentives for the incorporation of health and safety considerations in design decisions (Schulte et al., 2008).

In 2011, the American Society of Safety Engineers (ASSE) announced the approval of the American National Standards Institute (ANSI)/ASSE Z590.3 standard, 'Prevention through Design: Guidelines for Addressing Occupational Risks in Design and Redesign Processes'. The standard provides guidance on including PtD concept within an occupational safety and health management system. It further provides guidance for a lifecycle assessment and design model that balances environmental and occupational safety and health goals over the life span of a facility, process, or product (American Society of Safety Engineers, 2011). Another PtD initiative is the Design for Construction Safety which is the process of addressing construction site safety and health in the design of a construction project. This initiative is the product of the US OSHA Alliance Program's Construction Roundtable, a platform which allows OSHA partici-pants who share a common interest in construction related topics and issues to discuss and share experiences on workplace health and safety (Toole, 2014).

In Singapore

The Singaporean construction industry recognises the importance of incor-porating safety into a project early in the design stage. In collaboration with the Ministry of Manpower, the Singapore Workplace Safety and Health (WSH) Council established the Design for Safety (DfS) initiative. DfS is a concept that aims to change the construction value chain culture by bringing clients and designers into the 'safety picture' as early as the concept design stage to improve safety and to reduce rework through effective project planning (WSH Council, 2012). In 2008, the WSH Council launched the Guidelines on DfS in Buildings and Structures to assist the clients, designers, and contractors in the process of design safety and the transfer of vital safety and health information along the construction process chain. The guidelines specify duties of designers which include assessing the design to review risks that the design creates, eliminating hazards as far as reasonably practicable and providing mitigation strategies for remaining hazards. In short, designers should understand how the building or structure can be constructed, cleaned, maintained and decommissioned safely. Designers, including professional engineers and architects, should have relevant qualifications as required by the professional bodies in their discipline and should also have safety and health experience. In addition, the guidelines introduced a new role, the DfS Coordinator, who is appointed by the client to help facilitate DfS (WSH Council, 2011).

The above sections have discussed safety in design in four countries, namely, the UK, Australia, the USA and Singapore. In our opinion, there should be more of such national/state policies in all countries and states. Generally speak-ing safety in design policies (either explicit or implicit) is more in evidence in developed countries, while less is found in developing countries. Therefore, the above-mentioned four countries may represent the current best practice bench-mark on this topic.

Safety risk management

A number of different approaches and tools have been developed so that safety risks can be identified either during the designing process or via a design review process. These processes include design reviews and checklists used to identify, assess and mitigate safety risks in a design. Designers and engineers in charge of designing should include safety as one of the key tasks and objectives during design along with aesthetics and functionality as the brief (Hinze & Wiegand, 1992). In this section we focus on risk management as a method for safety in design. We discuss the process of risk management in the context of safety risk identification, assessment and mitigation at the design stage; we also discuss the related concepts and requirements, such as lifecycle risk analysis, knowledge and capability required to undertake safety risk management and information and documentation transfer. We base our discussions in the context of safety risk management at the design stage by referring to the commonly accepted standards and guidelines, such as AS/NZS ISO 31000:2009 Risk Management – Principles and Guidelines (2009), SA/SNZ HB 436:2013 Risk Management Guidelines – Companion to AS/NZS ISO 31000:2009 (2013), and HB 205-2004 OHS Risk Management Handbook (2004).

What is safety risk?

The concept of risk is applicable to nearly every human decision-making action of which the consequences are uncertain. This uncertainty arises because a key characteristic of decision making is its orientation towards the future. Nobody can, of course, ascertain the future. From an organisational perspective, there are numerous internal and external factors that influence the operations of organisations. This also causes an uncertain environment in which organisations are not certain whether, when, and the extent to which they will meet or exceed their objectives. The impact of this uncertainty on the organisation's objectives is considered as 'risk' (Australian Standards & New Zealand Standards, 2009). Another aspect of a risk which we should acknowledge is that risk and reward normally go hand in hand. For example, by agreeing to participate in a project, a contractor is taking a risk, but also has an opportunity to make a profit from the project. In the context of safety management, a safety risk can be defined as the likelihood and consequence of a potential injury or harm occurring (Standards Australia, 2004).

Safety risk management process

Risk management has become an important management tool due to the potential detrimental impact of a risk on businesses. British Standards BS 6079-3:2000 defines risk management as a systematic application of policies, procedures, methods, and practices to the tasks of identifying, analysing, evaluating, treating and monitoring risks (British Standards Institution, 2000). Australian and New Zealand Standards AS/NZS ISO31000:2009 defines risk management as coordinated activities to direct and control an organisation with regard to risk, and it includes risk management policy, principle, framework and process (Australian Standards & New Zealand Standards, 2009). The aim of risk management is not to avoid risks, but to make informed decisions, avoid unpleasant surprises, identify opportunities and encourage people to think more carefully about the consequences of their decisions and to ensure that an organisation's objectives are achieved (McGeorge & Zou, 2013).

The AS/NZS ISO31000 (2009) risk management standard outlines risk management process includes establishing the context, risk identification, risk analysis, risk evaluation and risk treatment; aparallel to these steps are communication and consultation, and monitoring and review. A basic process of safety risk management contains several steps as discussed in the following sections, which are based on the Australian Standards and New Zealand Standards (2009), Safe Work Australia (2011), and Zou et al. (2007):

1. *Risk communication and consultation*

    When implementing safety risk management, it is necessary to communicate and consult regularly with key stakeholders of the organisation or project, to obtain their support, perceptions, views and buy-in, and to utilise their expert knowledge. In the context of safety in design, this means that in the constructability or design review workshops, it is important to invite designers, clients, head contractors and subcontractors, and material and equipment suppliers to attend and participate. Consultation with relevant stakeholders should also occur when design changes are proposed, when new information becomes available and during each step of the risk management process. Lingard et al. (2013) emphasised the role of external stakeholders, which was often overlooked by proponents of safety in design. They stated that influential external stakeholders may have certain design preferences that are against design decisions that would reduce safety risk during construction. Therefore, early engagement of external stakeholders and open, trustworthy communication are essential for safety in design.

    Communication should be done in a way that is appropriate to each stakeholder because the needs of each stakeholder will be different in the detail and presentation of the information given (Standards Australia, 2004). It is also important to record all decisions made regarding safety risk assessment and mitigation strategies. At the end of the design stage, this information should be passed to contractors and facility managers.

    Consultation is a legal requirement and an essential part of managing work health and safety risks. It is easier to achieve a safe workplace when people involved at the design stage communicate with one another about potential risks and work together to find solutions. *So far as is reasonably practicable,* a designer should consult the client, other designers, project managers, and workers to make more informed decisions about how the structure/building can be designed to eliminate or minimise risks (Safe Work Australia, 2012c). Consultation also helps to develop a shared vision, shared objectives and shared responsibilities among stakeholders.

2. *Establishing the context*

In this step, the organisation or project team should define and explain its objectives in relation to safety performance and its indicators as well as the measurement methods, define the external and internal parameters to be taken into account when managing safety risk and determine the scope and risk criteria. The external context is the external environment, such as sociocultural, political, economic, technological and natural environments at the international, national, regional, or local level, where the organisation intends to achieve its objectives. Understanding the external context ensures that the expectations of external stakeholders are considered when developing objectives and risk criteria. The internal context is anything within the organisation or project that can influence the way in which safety risk is managed. This signifies that the safety risk management process should be aligned with the organisation's culture, processes, structure, strategies and work practices.

Thereafter, based on the context, the organisation should specify how risk management will be applied to manage workplace safety, the resources required, responsibilities and authorities and the records to be kept. There should be a defined structure which involves separating the activity, project, or change, into a set of elements or steps to provide a logical framework that helps ensure that significant safety risks are not overlooked (Standards Australia, 2004). The organisation should also define the safety risk criteria to evaluate the significance of risk, including identifying and measuring the consequences of risk, defining likelihood, and determining the level of risk and/or combination of risks (Australian Standards & New Zealand Standards, 2009). The criteria are used to decide whether a safety risk needs to be treated and the priorities for its treatment. The criteria must take account of legislative requirements to eliminate hazards and, where hazards cannot be eliminated, to minimise safety risks as far as practicable. The criteria should help decision makers to decide whether a risk has been reduced to an acceptable level and whether more treatment is practicable. Besides protecting the safety of people in the workplace, there may be other objectives, such as minimising financial losses to the organisation, which follow from safety problems. These may provide additional criteria to be considered (Standards Australia, 2004).

3. *Risk identification*

This is the process of identifying which risks may affect the safety objectives and documenting their characteristics. The risk identification process should be systematic and include not only what the risks are, but also where, when, why and how they might happen (McGeorge & Zou, 2013). In designing for safety, designers (and other relevant stakeholders) should identify foreseeable hazards associated with the design of a structure or facility (Safe Work Australia, 2012c). It is important during the design review workshop to identify safety risks, to assess high-risk issues, and where changes are planned (Standards Australia, 2004). Hazards may arise from the following aspects of work and their interactions: physical work environment; equipment, materials, and substances used; work tasks and how they are performed and work design and management.

There are many methods that can be used for identifying safety risks as specified in the standards and guidelines, such as AS/NZS ISO 31000 (Australian Standards & New Zealand Standards, 2009), HB436 (Standards Australia & Standards New Zealand, 2013) and HB205 (Standards Australia, 2004). Many of these methods are useful for identifying design-related safety risks and we suggest that these methods are considered and used whenever and wherever suitable, particularly the team-based risk identification methods. Some methods to identify hazards and safety risks in the design stage are:

- Consulting workers or people who are familiar with the construction process. Asking workers (face-to-face or via surveys) about safety problems that they have encountered in doing similar work and any near misses or incidents that have not been reported can be a useful way to identify hazards at the worksite level (Safe Work Australia, 2011).
- Reviewing available information. This is perhaps the most common method of hazard and risk identification. Information and advice about hazards and risks relevant to particular industries and types of work are available from regulators, industry associations, unions, technical specialists and safety consultants. Manufacturers and suppliers can also provide information about hazards and safety precautions for specific substances (safety data sheets), plant or processes (instruction manuals) (Safe Work Australia, 2011).
- Reviewing existing organisation records. Some organisations keep a risk register to help their employees identify risks in a particular task or process. Past audit reviews, accident reports and safety committees' reports are other sources of information to identify safety risks (Standards Australia, 2004).
- Inspection checklists. Generic checklists developed on the basis of safety issues identified in different types of workplace can be useful to show things that might go wrong and how hazards might arise when observing work processes. Checklists, however, must be used with care as they may not cover specific safety issues in a particular workplace (Standards Australia, 2004).
- Constructability reviews. Constructability involves the incorporation of construction knowledge in the design of a structure. A high level of constructability means that the structure is not exceedingly difficult to build, so the project will not be overly expensive and can be built within a timeframe. A constructability review process involves periodic reviews throughout the design stage to ensure that a project meets the

intended level of constructability (Gambatese, 2000). It is most useful when these reviews involve managers, designers, technical experts and people who are experienced in construction processes. The broader the experience represented, the more hazards and risks will be identified (Standards Australia, 2004).

4. *Risk analysis and evaluation*

The outcome of the risk identification process is a list of hazards and scenarios that could represent a threat to project safety objectives. These need to be analysed to understand their potential impact on the safety objectives so that they can be ranked and prioritised. Risk analysis aims to determine risk magnitude which is a combination of a risk's consequence and its likelihood. By finding their magnitude, critical risks can be detected and controlled (McGeorge & Zou, 2013). The actions that should be taken to perform safety risk analysis are (Safe Work Australia, 2011):

- Determining how severe the harm could be, by considering the following questions: What type of harm could occur? How severe would the harm be (death, serious injuries or minor injuries)? What factors could influence the severity of harm that occurs (the distance someone might fall, the concentration of a particular substance and whether the harm may occur immediately or may take time for it to become apparent)? How many people are exposed to the hazard within and outside the workplace? Could one failure lead to other failures? Could a small event escalate to a much larger event?
- Determining how hazards may cause harm. Many incidents occur as a result of a chain of events and a failure of one or more links in that chain. If one or more of the events can be stopped or changed, the risk may be eliminated or reduced. Some factors to consider are: the effectiveness of existing control measures, how work is actually done rather than relying on written manuals and procedures and infrequent or abnormal situations.
- Determining the likelihood of harm occurring. This can be estimated by considering the following questions: How often is the task done and does this make the harm more or less likely? How often are people near the hazard and how close do people get to it? Has it ever happened before and how often?

Risks can be quantified by the following formula:

Safety risk magnitude

= likelihood of occurrence × severity of resulting harm

A meaningful numerical value should be assigned to both the likelihood and severity aspects of the formula. Tables 6.1 and 6.2 give examples to this evaluation process by using a five-point scale format. The severity of harm can be measured by using different indicators, such as the impact on the

Table 6.1 Evaluating likelihood of occurrence

Value	Descriptor	Examples
1	Rare	May occur only in exceptional circumstances
2	Unlikely	Small likelihood, 1 in 100 or less
3	Occasional	Might occur at some time
4	Likely	Will probably occur in most circumstances
5	Frequent	Is expected to occur in most circumstances

Table 6.2 Evaluating severity of harm

Value	Descriptor	Examples
1	Insignificant	No injuries, low financial loss
2	Minor	Minor injury, first aid treatment, on-site release, immediately contained, medium financial loss
3	Moderate	Reportable injury or illness, medical treatment required, on-site release, contained with outside help, high financial loss
4	Major	Major injury or illness, loss of production capability, off-site release with no detrimental effects, major financial loss
5	Catastrophic	Fatality, toxic release off-site with detrimental effects, huge financial loss

worker compensation, time delay and cost increase of a project (Howarth & Watson, 2009; Standards Australia, 2004).

Table 6.3 presents the risk magnitude matrix resulting from the calculation using the above formula and the required actions based on the risk rating (Howarth & Watson, 2009).

5. *Risk treatment, response and control*

Having identified and analysed the risks, the next step is risk response and control, which is an action or a series of actions designed to deal with the presence of risk (McGeorge & Zou, 2013). In this step, the level of risk determined in the analysis process is compared against criteria of whether to treat the risk. The options are: no action is required, the risk is monitored, the risk level must be reduced using existing knowledge, the design must be changed, and further analysis is required to define the best controls (Standards Australia, 2004).

A hierarchy of control should be considered when reducing and controlling safety risks. The hierarchy presents the ways of controlling safety risks ranked from the highest level of protection and reliability to the lowest. There are six control measures in the hierarchy, as follows (Safe Work Australia, 2011):

- *Eliminate hazards.* In using this hierarchy, designers and other stakeholders involved in the risk management process should always aim to eliminate a hazard. If this is not reasonably practical, then the risk

Table 6.3   Risk rating matrix

Likelihood rating	Severity rating				
	1	2	3	4	5
5	5	10	15	20	25
4	4	8	12	16	20
3	3	6	9	12	15
2	2	4	6	8	10
1	1	2	3	4	5

Note:

      = low risk rating
      = medium risk rating
      = high risk rating

For low rate risks:
- Check that no further risks can be eliminated by modifying design, construction or method of work
- Check that no further cost effective control measures can be applied
- Manage by routine procedures

For medium rate risks:
- Consider an alternative design, construction or method of work
- Utilise additional control measures, precautions to be adopted
- Designers must communicate these remaining or residual hazards to contractors

For high rate risks:
- Seek an alternative design, construction or method of work
- Specify control measures to reduce risk rating to acceptable level
- Designers must communicate these remaining or residual hazards to contractors

should be minimised by working through the other alternatives in the hierarchy. Eliminating hazards and associated safety risks is the most effective control measure. This method is often cheaper and more practical to achieve at the planning or design stage of a project, when there is a greater scope to design out hazards or incorporate risk control measures that are compatible with the original design and functional requirements. For example, a designer can remove trip hazards on the floor.

- *Substitute the hazard with something safer.* This means designers should consider using structural forms, materials or methods that are safer for construction and maintenance. For example, a designer can substitute solvent-based paints with water-based ones. Besides being easy to apply and clean, water-based paints also emit low amounts of volatile organic compounds, which improves the indoor air quality.

- *Isolate the hazard from people.* This involves physically separating the source of harm from people by distance or using barriers. A common example of this method is installing guardrails around exposed edges and holes in floors.
- *Use engineering controls.* An engineering control is a control measure that is physical in nature, including a mechanical device or process. Some examples are using mechanical devices to move heavy objects, placing guards around moving parts of machinery, and installing electrical safety switches.
- *Use administrative controls.* These are work methods or procedures designed to minimise exposure to a hazard. For example, developing safe work procedures and safe work method statement, and using signs to warn people of a hazard.
- *Use personal protective equipment (PPE).* These include wearing hard hats, safety shoes, gloves and protective eyewear.

Elimination and substitution are generally considered the most effective form of risk control and they are more relevant to safety in design. It should be noted that administrative controls and PPE should only be used when there are no other practical control measures available, as an interim measure until a more effective way of controlling the safety risk can be used, and as a back-up to supplement higher level control measures. The costs and benefits of the control measures should be assessed. Chapter 2 provides a process and a case study for this purpose.

6. *Risk monitoring and review*

   The control measures that are put in place should be reviewed regularly to make sure that they work as planned. A review is particularly required (Safe Work Australia, 2011)

   - when the control measure is deemed ineffective;
   - before a change at the work place that is likely to introduce a new or different risk for which the control measure may not be effective;
   - if a new hazard or risk is identified;
   - if the results of consultation or communication with other stakeholders indicate that a review is necessary;
   - if a health and safety representative requests a review.

The following are examples of indicators that can be used to review and assess risk management activities (Standards Australia, 2004):

- The scope of the risk management process is adequately defined.
- Proper context where safety risk is managed has been established.
- The people involved have an appropriate range of expertise.
- There is evidence of communication with those who have an understanding of the risks.

- There is evidence of actively seeking what might happen, not only what has happened.
- An appropriate range of information sources is used and collected to identify hazards and risks.
- Evidence-based judgement is used to assess consequences and their likelihood.
- Any ranking tool is applicable to the scope and context situation.
- Recommended controls are high in the hierarchy.
- Residuals risks have been evaluated.
- Controls have been implemented and their effectiveness monitored.
- At the end of the review process, lessons learnt and key success factors are documented for future project references.

Lifecycle safety risk analysis

It should be noted that the risk management process described above is iterative, meaning it may be repeated many times with additional or modified risk evaluation criteria, thus leading to a process of continual improvement. It is essential that each step of the risk management process be documented, including assumptions, methods, data sources and results, to ensure traceability (Standards Australia, 2004).

Furthermore, there could be tension or conflict between solutions along the project lifecycle. For example, Lingard et al. (2013) investigated the outcome and relationship between implementing the safety in design concept for improving safety in construction and operational stages. They found tensions between construction and operational safety, meaning design decisions taken to reduce safety risk during the operational stage result in increased risk during construction. Therefore, they suggested that procedures and guidelines on implementing safety in design should provide practical guidance about how to identify and manage conflict and trade-offs in reducing safety risk across the lifecycle of a project. The use of risk management principles to implement safety in design assumes that design is stable at an early stage and that all foreseeable hazards can be identified and managed through risk management protocols. In reality, however, design is inherently uncertain and there may be changes to the design in the later stage of the project after the risk management exercise in the early design stage. This risk management process, therefore, should remain 'live' and be updated throughout the project lifecycle (Lingard et al., 2013).

Designers should consider how their design will affect the health and safety of people who will interact with the structure throughout its life. This means that designers should consider design solutions for reasonably foreseeable hazards that may occur as the structure is constructed, commissioned, used, maintained, modified, decommissioned, demolished and disposed (Safe Work Australia, 2012c; Schulte et al., 2008). Safe Work Australia (2012c) recommends a system approach to integrate the risk management process in the design stage and to encourage collaboration between key stakeholders, including the client, designer and contractor as follows:

1. *Pre-design phase.* Intended use of the building or structure should be determined by the client and designer. Information on relevant laws and regulations, industry statistics regarding injuries and incidents and guidance on potential hazards and possible solutions should be gathered from various sources to assist in identifying hazards, as well as assessing and controlling risks.

2. *Conceptual and schematic design phase.* Hazard identification should take place as early as possible in this phase. Designers and relevant stakeholders should determine hazards that can be affected, introduced, or increased by the design of the structure. At this phase, consideration should also be given to possible ways that hazards could be eliminated or minimised. The following are aspects which should be focused on during the preliminary hazard identification (Safe Work Australia, 2012c):

    • Siting of structure: proximity to adjacent property or nearby roads, surrounding land use, clearances required for construction equipment and methods, demolition of existing structures, proximity to underground and overhead services, site conditions and public safety.

    • High consequence hazards: dangerous goods, high energy hazards and health hazards.

    • Systems of work (involving the interaction of people with the structure): construction methods, construction materials, coordination and relationships among work activities, pedestrian and vehicle separation, cleaning and maintenance access, hazardous manual tasks, working at height and misuse throughout the structure lifecycle.

    • Environmental conditions: adverse natural events and disasters, inadequate ventilation or lighting, noise levels and welfare facilities.

    • Incident mitigation: the possibility of the structure to exacerbate the consequences after an incident due to inadequate egress, inconvenient location of assembly areas and inadequate emergency service access.

3. *Design development phase.* This phase involves the development of detailed drawings and specifications. Risk management activities in this phase are (Safe Work Australia, 2012c):

    • Developing a set of design options in accordance with the hierarchy of control.

    • Selecting the optimum solution by balancing the direct and indirect costs of implementing the design against the benefits derived.

    • Testing and evaluating the design solution.

    • Redesigning to control any residual risks.

    • Finalising the design, preparing a safety report and other risk control information needed for the structure's lifecycle.

Knowledge required

Besides the core design capabilities relevant to the designer's role, in order to implement safety in design effectively, a designer should also have (Safe Work Australia, 2012c):

- Knowledge of work health and safety legislation, codes of practice and other regulatory requirements.
- An understanding of the intended purpose of the structure.
- Knowledge of risk management processes, as discussed in the previous sections.
- Knowledge of technical design standards.
- Knowledge of material properties and their impact on human safety and health.
- An appreciation of construction methods and their impact on the design.
- An understanding of human behaviours when using and maintaining the structure.

Documentation and information transfer

Designers should record and transfer crucial information about hazards identified and action taken or required to control risks to those involved in later stages of the structure lifecycle. This communication is very important to inform others about any residual risks and to minimise the likelihood of safety features incorporated into the design being altered or removed in the subsequent stages of the structure lifecycle. The transfer of information can be done via a safety report which designers should prepare. The report should include information about any hazardous materials or structural features and the designer's assessment of the risk of injury or illnesses to construction workers arising from those hazards. The action that the designer has taken to control those risks should also be stated in the report. Rigorous documentation should be maintained by designers throughout the structure lifecycle to demonstrate their safety in design considerations. The documentation should include the safety report, risk register, safety data sheets, manual and procedures for safe maintenance, dismantling and demolition (Safe Work Australia, 2012c).

Current issues and possible solutions

Barriers for implementing safety in design

Despite the importance and benefits of safety in design, there are still barriers in the consideration of safety risks at the design stage. In some countries, including the USA, the involvement of designers in construction safety is voluntary. This may potentially create some barriers that hinder the implementation of safety in design concept as argued by Behm (2005) who observed this condition in the

US construction industry. The first barrier is legal and liability issues. Designers are wary of the increased liability, thus only a small number of designers are taking the lead in designing for construction safety. The legal and insurance system, for example, has caused architects to be afraid of getting involved in safety. Many designers also commented that legal counsel specifically advised them not to address construction worker safety in their design, particularly as there is no obligation to do so. The second barrier is the regulatory action, which strengthens the resistance to change caused by the first barrier. In the USA, in 1988, two bills that would have placed increased safety responsibility on designers were rejected due to opposition from a large segment of the construction industry. In 1999, another bill that would put a requirement in the State Building Code to design and install permanent anchor points on all buildings to minimise the number of falls to construction workers, maintenance workers and homeowners was also rejected due to concerns regarding the effectiveness of such anchor points. The third barrier is the traditional or design-bid-build construction procurement which separates the design and construction processes. As a result, contractors have no constructability input during the design stage, while contracts between owners and contractors always require contractors to be fully and solely responsible for on-site safety.

Designers' safety knowledge

Some designers do not acknowledge the relevance of their role in safety and some have deliberately avoided addressing construction safety to minimise their liability exposure. Although some admit that their designs impact safety performance, they argue that they do not know how to change their designs to improve or ensure safety (Gambatese & Hinze, 1999; Zhou et al., 2012). A central body of knowledge available for designers to address safety in their designs may be able to address this problem because such a system allows design knowledge to be accumulated and stored in a central location for all designers to access in their subsequent projects (Gambatese & Hinze, 1999). For example, Gambatese et al. (1997) collated more than 400 design suggestions for safety through literature searches, interviews with construction practitioners and reviews of worker safety manuals and safety design manuals. These suggestions were compiled in the 'Design for Construction Safety Toolbox'. This safety in design tool aims to introduce a variety of project-specific design suggestions that would improve safety during the construction stage. The tool allows users to customise the process by entering initial information about the project and the users' design discipline. The following are examples of the safety in design suggestions recorded (Gambatese et al., 1997):

- Design components to facilitate prefabrication in the shop or on the ground so that they may be erected in place as complete assemblies. *Purpose:* Reduce worker exposure to falls from elevation and the risk of workers being struck by falling objects.

- Design steel columns with holes in the web at 0.53 and 1.07 m above the floor level to provide support locations for guardrails and lifelines. *Purpose:* By eliminating the need to connect special guardrail or lifeline connections, such fabrication details will facilitate worker safety immediately upon erection of the columns.
- Design underground utilities to be placed using trenchless technologies. *Purpose:* Eliminate the safety hazards associated with trenching, especially around roads and pedestrian traffic surfaces.
- Route piping lines that carry liquids below electrical cable trays. *Purpose:* Prevent the chance of electrical shock due to leaking pipes.

Design-build contracts

Designers who are working in design-build companies have better opportunities to address safety in their designs because they are able to work with their colleagues who are responsible for the construction stage of a project (Gambatese & Hinze, 1999). As a result, communication between designers and contractors improves, while good ideas are remembered and used on subsequent projects (Zhou et al., 2012). This echoes the importance of constructability reviews, in which safety in design should be part of the constructability review process. The separation of the design and construction stages is unique to the industry and constructability attempts to address this so that the design of a structure facilitates ease of construction. Effective communications between project stakeholders, particularly the client, designer and main contractor, are key to the successful implementation of constructability. These communications should be established and nurtured early in the design stage because decisions taken at this stage have a greater potential to influence the final outcome of the project than those taken in the later stages (McGeorge & Zou, 2013).

Case studies

Case 1 – risk and opportunity at design

Zou et al. (2008) undertook a case study on risk and opportunity at design (ROAD) by using a multi-national project and construction management organisation that implements safety in design through its ROAD programme, which is a company's compulsory process in every project. The ROAD process provides a forum for all key project participants to be involved in the process of hazard identification at the design stage. It promotes a sense of ownership by giving open lines of communication between stakeholders and by encouraging enthusiastic participation in the successful implementation of their own suggestions. The overall ROAD process is outlined in the following eight steps:

1. Assessing building elements using ROAD hazard/opportunity checklists.
2. Assessing trade packages using ROAD hazard/opportunity checklists.

3. Recording and uploading the ROAD document into the project management plan.
4. Including ROAD items as part of the agenda in design meetings.
5. Establishing ROAD action and status lists.
6. Updating and reporting ROAD status at each design review.
7. Considering actions drawn from the ROAD meetings prior to approval for
   construction.
8. Updating monthly the ROAD document as part of the management and
   project reviews.

In their study, Zou et al. (2008) interviewed several construction managers to
understand the effectiveness of ROAD in practice, and the interviewees claimed
that ROAD offers the following advantages (Zou et al., 2008):

- Constant identification of construction procedures ensures that the mitigation of risks is identified and responsibility is taken for them.
- ROAD creates a system of accountability and transparency within the construction delivery team.
- Stakeholders have the opportunity to contribute to ROAD and bring their
  knowledge from previous projects to inform the present one. This participation means that the stakeholders have the sense of ownership concerning
  their safety and the safety of other people.
- ROAD facilitates a critical analysis of the construction process and the constructability of the project.
- Risks and opportunities are identified early, allowing time and budgetary
  constraints to be adjusted.
- ROAD is on-going and evolves with the project. This flexibility ensures that
  safety measures can be accounted for throughout the delivery process.

Zou et al. (2008) also found that despite its advantages, ROAD does take time
and effort in its implementation. Some people may feel that ROAD is an extra
and unnecessary task. As such, the commitment from the top management
is crucial for its success. Project managers, design managers and construction
managers need to have the right attitude and motivation in leading the ROAD
process. The participation of experienced project managers, construction managers and designers, as well as head contractors and main subcontractors, is also
essential to highlight key risks and potential opportunities (Zou et al., 2008).

Case 2 – life cycle safety analysis

The case study presented in this section is drawn from Weinstein et al. (2005).
The life cycle safety (LCS) safety-in-design programme is implemented during
the design and construction of a $1.5 billion semiconductor manufacturing and
research facility in the USA (Weinstein et al., 2005). The LCS considers safety
concerns in all phases of the facility's lifecycle, including programming, detailed
design, construction, operations, maintenance, retrofit and decommissioning.

The programme was envisioned to be a comprehensive review process which not only involves the design firm, but also the client, the general contractor and numerous trade contractors. Initially, a safety-in-design task force, which consisted of representatives of the client, the design firm and the contractor, was formed. A third-party consultant also participated in the task force as the facilitator of the process. Thereafter, seven discipline-based workgroups evaluated design options against two similar previous projects and reported their recommendations to the task force. The task force assessed these recommendations on the basis of certain criteria, including cost, schedule, environmental sustainability, adaptability to future manufacturing technologies, and improved safety through design. During the detail design phase, trade contractors also provided input on 22 different design packages. The involvement of the main contractor and participating trade contractors is valuable in the LCS.

The following techniques were used throughout the LCS process to enhance safety:

- *Safety-in-design checklist.* The design firm developed a 101-item safety-in-design checklist based on lessons learnt from previous projects. The items consisted of design issues which might cause safety problems during the construction and operation of the facility.
- *Focused group interviews.* The LCS taskforce organised six focused group interviews early in the design stage with several trade contractors, client facility technicians and vendor tool technicians who had worked on other similar projects. Analysis of interview transcripts identified 196 distinct comments related to safety in design.
- *LCS review comments.* Client maintenance technicians, trade contractors and environmental safety and health staff were involved in the LCS review, which focused on issues related to safe design. This was a detailed design review process of the 22 design packages which produced 789 design review comments.
- *Technical review comments.* These regular reviews between the client and designer personnel were aimed to verify and improve the technical characteristics and qualities of the design. In the project, 7071 review comments were made.
- *Exit focus group interviews.* The LCS taskforce conducted 29 exit group interviews with the general foremen, site superintendents and safety personnel of trade contractors after they completed their work to identify various safety-in-design issues that they faced. They represented over 90% of the construction man-hours on the project.

There are principles of safety in design which can be learnt from the LCS case:

- A safety-in-design concept should ideally be implemented as early as possible during the design stage because it is easier and cheaper to influence safety that way.

- The project proved that early interaction between designers, contractors and other relevant stakeholders can result in safer designs. This process can be easily introduced within a design and build project delivery system, but it is not impossible to introduce such techniques within a traditional design-bid-build project delivery system. Contractors who participate in design reviews in the design stage may see this as an opportunity to learn more about the project, thus increasing their possibility to win work in the project.
- Maintaining proper documentation is valuable for information transfer purposes, not only from the designer to the contractor and, finally to the client, but also for developing a safety-in-design checklist which can be used in future projects.

Building information modelling (BIM) for safety in design

BIM, including a range of digital tools, such as online databases, virtual reality, geographic information systems (GIS), 4D (or nD) CAD, and location-based sensing and warning technologies have a potential to change the way safety can be approached by automatically detecting and eliminating hazards. As discussed in the previous sections, safety in design is mainly about the identification of potential hazards and the decision of choosing corresponding safety measures in the design stage. As such, accurate and precise identification of potential safety hazards is critical to the *safety in design* process. In practice, failures in identifying hazard are often due to limited experience, poor training and oversight of construction staff. Another issue is the separation between safety and design processes which may involve different actors who do not communicate sufficiently with one another. This issue creates difficulties for safety personnel to analyse what, when, why, and where safety measures are needed. BIM has a potential to change the way safety can be approached by automatically detecting and eliminating hazards, either at the design or construction stage (Zhang et al., 2013). For example, BIM allows greater details to be developed earlier in a project. This may enhance designers' awareness of construction safety issues (Zhou et al., 2012). A range of digital tools, as listed earlier, has been developed and used to help contractors manage safety in the design and construction stages. However, in comparison, digital tools for managing safety in design are less mature and relatively limited in their application (Zhou et al., 2012), and in the following sections, we discuss several BIM-based examples that are directly or indirectly related to safety in design.

One example of using BIM for safety in design is the ToolSHed™ (Cooke et al., 2008), which is an information and decision support tool to help designers integrate the management of safety risks into the design process. ToolSHed™ is a web-based tool developed to provide designers with specialist safety knowledge and guidance. Knowledge was acquired from Australian

Occupational Health and Safety guidance material, industry standards and codes, and an expert panel. The knowledge was modelled in a series of logic diagrams which represent a template for reasoning in complex situations. At present, ToolSHed™ only deals with the design-related risks of falling from heights during maintenance work on building roofs. The risk assessment prompts designers to enter information about relevant design features that could impact upon the risk of falling from a height. A risk report is generated to advise the designers about the level of risk of falling from heights, and an explanation of the design factors contributing to the inferred level of risk (Cooke et al., 2008).

The second example is Zhang et al. (2013) who developed an *automated* rule-based safety checking system for fall prevention due to openings in slabs, edges on floor, and openings in walls. An initial set of rules was generated using a set of fall-prevention rules for the three conditions. Once the rules were established, the system was able to detect various locations requiring fall protection, based on design drawings; for example, exterior walls are examined to determine where edge protection is needed; openings in slabs are examined to prevent fall through openings; openings in exterior walls are examined to determine where additional wall opening protection is required and interior walls around slab openings are examined for fall protection from wall openings. This tool can be used in the design stage to identify fall hazards during different stages of construction.

As the third example, BIM is also valuable during constructability reviews, in which many argue that they are able to improve construction safety. For example, in a project, designers may have design BIM models, the contractor has a BIM model for use in sequencing the work and major suppliers have prefabrication BIM models of building elements. An integrated BIM model with information derived from these different disciplines will help decision-makers to identify, visualise and resolve conflicts among various building systems when conducting constructability reviews (Sullivan, 2007). As discussed earlier, projects with high levels of constructability facilitate ease of construction which promotes better safety performance.

Sulankivi et al. (2014) use BIM to promote constructability which has positive impacts on construction site safety. A basic prerequisite of good constructability is the integration of various elements in correct and accurate design drawings. BIM can assist in this integration process by combining drawings from various design disciplines into one file and performing semi-automatic clash detection. This way, design conflicts, for example, between structural elements and mechanical components, can be identified and eliminated. Fewer design conflicts and errors lead to fewer disruptions on site and less *ad hoc* decisions, which are known to increase safety hazards. BIM also can become a useful tool for cooperation between designers and contractors to improve constructability and safety. Detailed BIM models have been found to improve the visualisation of constructability issues, thus becoming a useful tool during constructability assessment meetings.

Conclusions

In this chapter, we have discussed the concept of safety in design, which fundamentally means the integration of control mechanisms to eliminate or minimise safety risks at the design stage. This concept is important because research has revealed that many incidents can be traced back to be design-related. Furthermore, safety in design is needed to eliminate safety risks during the design stage as it is cheaper and easier to implement than making changes later, when the risks become real in the workplace.

The chapter continued by discussing safety in design legal requirements, regulations and initiatives in the UK, Australia, the USA and Singapore. Although the concept is still not mandated by law in the USA and Singapore, it has gained popularity due to its potential benefits to ensure safe construction process and safe building operation. Despite this, barriers and issues to its implementation still exist. Designers may not have enough safety knowledge and hence be unwilling to take the liability for safety. The separation between the design stage and the construction stage also causes difficulties for designers and contractors to communicate and identify safety risks which may arise in the design stage.

The implementation of the safety risk management process is fundamental to overcome these barriers. The risk management process has been detailed in this chapter, together with case studies on the implementation of safety in design in the construction and engineering industry. The potential of BIM as a tool to implement the safety in design has also been explored. With the increasing penetration of information technology in the construction and engineering industries, BIM will become more and more popular and necessary. BIM, in time, will be increasingly applied for safety in design and constructability reviews. To conclude, safety in design is important and we encourage universities and industry professional bodies to provide sufficient training on the topics and issues discussed in this chapter to students and practitioners.

References

American Society of Safety Engineers (2011). ANSI/ASSE Z590.3-2011. *ASSE Tech Brief*. Retrieved 24 April, 2014, from https://www.asse.org/publications/standards/z590/docs/Z590.3TechBrief9-2011.pdf

Andres, R. N. (2002). Risk assessment & reduction: A look at the impact of ANSI B11.TR3. *Professional Safety, 47*(1), 20–26.

Australian Standards, & New Zealand Standards (2009). AS/NZS ISO 31000:2009 Risk Management - Principles and Guidelines. Sydney, Australia/Wellington, New Zealand: Standards Australia/Standards New Zealand.

Behm, M. (2005). Linking construction fatalities to the design for construction safety concept. *Safety Science, 43*(8), 589–611.

British Standards Institution (2000). *BS 6079-3:2000 Project Management Guide to the Management of Business Related Project Risk*. London: British Standards Institution.

CDM Regulations (2007). *Construction (Design and Management) Regulations 2007*. UK: UK Government.

Centers for Disease Control and Prevention (2013). Engineering education in occupational safety and health Retrieved 26 May, 2014, from http://www.cdc.gov/niosh/topics/shape/

Consultants' Health & Safety Forum (2012). *Safe by Design*. Bootle, UK: Health and Safety Executive.

Cooke, T., Lingard, H., Blismas, N., & Stranieri, A. (2008). ToolSHeD: The development and evaluation of a decision support tool for health and safety in construction design. *Engineering, Construction and Architectural Management, 15*(4), 336–351.

European Foundation. (1991). *From Drawing Board to Building Site*. London: Her Majesty's Stationary Office.

Gambatese, J. A. (2000). Safety constructability: Designer involvement in construction site safety. Paper presented at the Construction Congress VI, Orlando, USA.

Gambatese, J. A., & Hinze, J. (1999). Addressing construction worker safety in the design phase: Designing for construction worker safety. *Automation in Construction, 8*(6), 643–649.

Gambatese, J. A., Hinze, J., & Haas, C. T. (1997). Tool to design for construction worker safety. *Journal of Architectural Engineering, 3*(1), 32–41.

Hale, A., Kirwan, B., & Kjellén, U. (2007). Safe by design: Where are we now? *Safety Science, 45*(1–2), 305–327.

Health and Safety Executive (2014). General CDM 2007 Retrieved 19 March, 2014, from http://www.hse.gov.uk/construction/cdm/faq/general.htm

Hinze, J., & Wiegand, F. (1992). Role of designers in construction worker safety. *Journal of Construction Engineering and Management, 118*(4), 677–684.

Howarth, T., & Watson, P. (2009). *Construction Safety Management*. Chichester, UK: Wiley-Blackwell.

International Labour Office. (1992). *Safety and Health in Construction*. Geneva: International Labour Organization.

Lingard, H., Cooke, T., Blismas, N., & Wakefield, R. (2013). Prevention through design: Trade-offs in reducing occupational health and safety risk for the construction and operation of a facility. *Built Environment Project and Asset Management, 13*(1), 7–23.

McGeorge, D., & Zou, P. X. W. (2013). *Construction Management: New Directions* (3rd ed.). Chichester, UK: Wiley.

National Occupational Health and Safety Commission (2003). *Eliminating Hazards at the Design Stage (Safe Design): Options to Improve Occupational Health and Safety Outcomes in Australia*. Canberra: National Occupational Health and Safety Commission.

NIOSH. (2014). *The State of the National Initiative on Prevention through Design*. Cincinnati, OH: NIOSH.

Safe Work Australia (2011). *How to Manage Work Health and Safety Risks: Code of Practice*. Canberra: Safe Work Australia.

Safe Work Australia. (2012a). *The Australian Work Health and Safety Strategy 2012–2022*. Canberra: Safe Work Australia.

Safe Work Australia (2012b). *Guide to the Work Health and Safety Act*. Canberra: Safe Work Australia.

Safe Work Australia (2012c). *Safe Design of Structures: Code of Practice*. Canberra: Safe Work Australia.

Schulte, P. A., Rinehart, R., Okun, A., Geraci, C. L., & Heidel, D. S. (2008). National Prevention through Design (PtD) Initiative. *Journal of Safety Research, 39*(2), 115–121.

Standards Australia. (2004). *OHS Risk Management Handbook (HB 205-2004)*. Sydney, Australia: Standards Australia.

Standards Australia, & Standards New Zealand (2013). *SA/SNZ HB 436:2013 Risk Management Guidelines - Companion to AS/NZS ISO 31000:2009*. Sydney, Australia and Wellington, New Zealand: Standards Australia/Standards New Zealand.

Sulankivi, K., Tauriainen, M., & Kiviniemi, M. (2014). Safety aspect in contructability analysis with BIM. Paper presented at the CIB W099 Achieving Sustainable Construction Health and Safety, Lund, Sweden.

Sullivan, C. C. (2007). Integrated BIM and design review for safer, better buildings. *Architectural Record, 6*, 191–199.

Toole, T. M. (2014). Prevention through design: design for construction safety Retrieved 24 April, 2014, from http://www.designforconstructionsafety.org/index.shtml

Weinstein, M., Gambatese, J. A., & Hecker, S. (2005). Can design improve construction safety?: Assessing the impact of a collaborative safety-in-design process. *Journal of Construction Engineering and Management, 131*(10), 1125–1134.

WSH Council (2011). *Guidelines on Design for Safety in Buildings and Structures*. Singapore: WSH Council.

WSH Council (2012). Design for safety in buildings and structures Retrieved 24 April, 2014, from https://www.wshc.sg/wps/portal/DesignForSafetyInBuildingsAndStructures?openMenu=1

Zhang, S., Teizer, J., Lee, J.-K., Eastman, C. M., & Venugopal, M. (2013). Building information modeling (BIM) and safety: Automatic safety checking of construction model and schedules. *Automation in Construction, 29*, 183–195.

Zhou, W., Whyte, J., & Sacks, R. (2012). Construction safety and digital design: A review. *Automation in Construction, 22*, 102–111.

Zou, P. X. W., Redman, S., & Windon, S. (2008). Case studies on risk and opportunity at design stage of building projects in Australia: Focus on safety. *Architectural Engineering and Design Management, 4*(3–4), 221–238.

Zou, P. X. W., Yu, W. Y. X., & Sun, A. C. S. (2009). An investigation of the viability of assessment of safety risks at design of building facilities in Australia. Paper presented at the CIB W099 Conference on Planning, Designing, and Building a Healthy and Safe Construction Industry, Melbourne, Australia.

Zou, P. X. W., Zhang, G., & Wang, J. (2007). Understanding the key risks in construction projects in China. *International Journal of Project Management, 25*(6), 601–614.

7 Research Methodology and Research–Practice Nexus

This chapter discusses research methodologies and methods applied in safety research in construction and engineering. It also discusses the nexus between research and practice, and proposes a model to bridge the gap between the two, with an aim to further improve safety performance in construction and engineering projects. In any field of professional endeavour, it is important to consider and employ appropriate research methods, which are underpinned by sound reasoning, that is, methodological arguments, and which are directly applicable in practice. In the case of safety management in construction and engineering, it is essential that research results and outcomes, that is, the knowledge and understanding developed through the research, can be applied in or have implications for safer work practice. Additionally, research should stimulate continuous improvement and facilitate safety learning. This chapter addresses the following issues:

- Research methodologies commonly used in social science research, which can provide a framework for their adoption in construction safety research.
- The effect of social desirability bias (SDB) on safety research design, together with techniques to minimise SDB.
- The implications of prevailing methodologies and the nature of the knowledge which these methodologies generate, with particular reference to construction safety practices and performance.
- An alternative research–practice nexus paradigm is proposed with the intention of increasing collaboration between researchers and industry practitioners, thus improving safety performance in construction and engineering projects.

Strategic Safety Management in Construction and Engineering, First Edition.
Patrick X.W. Zou and Riza Yosia Sunindijo.
© 2015 John Wiley & Sons, Ltd. Published 2015 by John Wiley & Sons, Ltd.

A typical research process

Research essentially is an organised inquiry carried out to provide information for solving problems (Cooper & Schindler, 2008). It is a systematic process of collecting, analysing and interpreting data to increase understanding of a phenomenon. There are eight typical characteristics of research (Leedy & Ormrod, 2013):

- Research originates with a research question or a problem.
- Research requires clear articulation of a research goal, that is, the aim of the research.
- Research usually divides the principle problem into more manageable sub-problems.
- Research is guided by the specific research problem, question or hypothesis.
- Research requires a specific plan for proceeding.
- Research rests on certain critical theories.
- Research requires the collection and interpretation of data in an attempt to resolve the problem that initiated the research.
- Research is cyclical; there is no obvious end point because research encourages follow-up studies.

The research process is basically the steps that are undertaken to carry out the research from the beginning (determining research problems) until the end (reporting results). A typical research process is presented in Figure 7.1. It should be noted, however, that research is a dynamic process and so flexible in its implementation. All the steps need to be adapted as the research progresses and findings emerge, while forming all aspects into a coherent chain. Maintaining coherence and complementarity is essential to produce robust research results and conclusions (Fellows & Liu, 2008).

Depending on the nature of the research, the first step in the research process is generally to confirm the broad topic of the research and identify the research problem, which can be theoretical or applied or both. Research that focuses on theoretical problems, that is, pure or basic research, is mainly geared towards advancing knowledge, while applied research aims to provide solutions or recommendations to meet the needs of the industry. By and large, the latter is more relevant to safety research, which aims to bring about safety improvement in the industry.

Once the research problem is identified, research aims and objectives can be determined to address the problem. This step requires an overview of the existing literature to identify the research gap and to ensure that the research has not been carried out previously. The next step is a critical and comprehensive review of existing literature. It is essential to explore existing theories and previous research findings to examine what has been done, to clearly understand issues that have not been resolved, to identify challenges, and to avoid mistakes, particularly those that have happened in previous research. This review

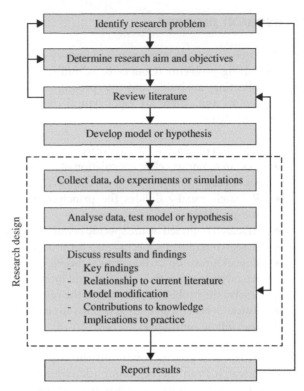

Figure 7.1   Typical research process

of literature may result in the need to further define the research problem and refine the research aim and objectives. In a deductive type of research, a conceptual research model or hypothesis is developed after the review of literature. In the later stages, the model or hypothesis is tested to see whether it is actually applicable in practice, while also confirming whether the hypothesis should be accepted or rejected.

The next step is research design for collecting and analysing data, which can be considered as the core of the research. Research design is essentially a plan to collect and analyse data to ensure that the research approach is appropriate in meeting the research aims and objectives and resolving the research problem. There are three basic steps of research design. First, researchers should collect necessary data or, depending on the aim of the research, perform experiments or simulations. Second, the data or the experimental and simulation results should be analysed. In some cases, this involves the testing of the conceptual model or hypothesis developed in the earlier stage. Third, the analysis results should be discussed and interpreted so as to make them meaningful in the context of the research. The discussion of results should summarise key findings, link back results to the existing literature to find similarities and differences and explain the implications of the findings on practice and knowledge development. Details on research design are discussed later in the chapter.

The last step of the research process is reporting the results. Without reporting, the research has no contribution. This reporting requires effective communication to ensure that the research realises its potential. A good research report explains the research process in a clear and concise manner. It starts by explaining the objectives and significance of the research. It then succinctly informs the readers of the theories and existing literature that are the foundation and stepping stones of the research. It shows a logical relationship among the objectives, methodology, design and methods of the research. The report continues by describing the analysis process and presenting the analysis results. It interprets the results and discusses the findings before concluding the report by showing its contributions on advancing the body of knowledge and/or providing recommendations to improve performance in industry practices. The research then can progress further by using the research process outlined in Figure 7.1.

Research methodologies

Research in safety in construction and engineering is typically classified as social research. Social research draws on the social sciences for theoretical inspiration and may be motivated by developments and changes in society (Bryman, 2012). In conducting social research, researchers should make clear the philosophical assumptions that they adopt because it will determine the appropriate research methodology to be employed and the type of knowledge that the research generates (Creswell, 2009; Dainty, 2008). There are two main philosophical assumptions in social research: ontology and epistemology, and we discuss each of them in detail in the following sections.

Ontological questions are concerned with the nature of social entities under investigation. There are two contrasting ontological positions: objectivism and constructivism. Objectivism asserts that social phenomena are independent of social actors, that is, they are beyond the reach or influence of these social actors. For example, a construction organisation has safety rules and regulations along with standardised procedures for getting things done safely. They act as constraining factors that compel workers to behave in certain ways, otherwise there may be consequences. In this case, safety becomes something external to social actors and has an objective reality. On the contrary, constructivism asserts that social actors continually accomplish social phenomena and their meanings. It implies that social phenomena are produced through social interaction and in a constant state of revision. Using the same example, constructivism views safety within an organisation as a negotiated order. It is worked at instead of having a pre-existing characteristic. As such, the social order in relation to safety is ever-changing because safety rules, regulations and procedures are continually being established, terminated or revised (Bryman, 2012; Burrell & Morgan, 1979).

Epistemological questions are concerned with the process of understanding the social phenomena and communicating the knowledge to others. It is about what should be regarded as acceptable knowledge in a discipline. Similar to ontological positions, there are also two contrasting epistemological positions: positivism and interpretivism. Positivism is an epistemological position that supports the use of the methods of natural sciences to the study of social phenomena. It views that knowledge can be gained in an objective way and that knowledge can be transferred in a tangible form. Interpretivism, on the other hand, considers that people and institutions, which are the subjects of social research, are fundamentally different from those of the natural sciences. Unlike atoms and molecules, a social phenomenon has a specific meaning and relevant structure for the people living, acting and thinking within it. According to interpretivism, knowledge is subjective and based on the experience and insight of individuals. As such, it has to be personally experienced rather than as something that can be acquired and conveniently transferred from one medium to another (Bryman, 2012; Burrell & Morgan, 1979).

The selection of the assumption to adopt fundamentally affects the selection of methodology (a general orientation to the conduct of social research), design (a plan to collect and analyse data), and methods (instruments to collect the data) of the research. For example, a research study which considers safety as an objective reality. This ontological position entails the use of positivism epistemological assumption to advance safety knowledge in an objective manner. Based on these assumptions, a quantitative research methodology is chosen to obtain numerical, standardised data that can be analysed statistically and generalised. A questionnaire is then developed as an instrument to collect the required data.

There are three common research methodologies adopted in social research: quantitative, qualitative, and mixed methods. Each methodology has its merits and shortcomings along with its supporters and critics, as discussed below.

Quantitative research

Quantitative research represents the dominant methodology in social research (Bryman, 2012; Leedy & Ormrod, 2013). This methodology typically tries to measure variables in a numerical way by using standardised instruments with a purpose to establish relationships among variables. The process involves the determination of concepts, variables and hypotheses at the beginning of the research, which are tested after data have been collected. The data themselves are collected from a population or from samples that represent the population so that research findings are generalisable (Leedy & Ormrod, 2013). Because quantitative research has an objectivist notion of social reality, researchers use established guidelines to conduct the research and try to remain detached from the phenomena and participants that they investigate to draw unbiased conclusions (Bryman, 2012; Leedy & Ormrod, 2013).

There are two primary research designs for conducting quantitative research (Creswell, 2009):

- *Surveys*, which provide a numeric description of trends, attitudes or opinions of a population by studying a sample of that population (Creswell, 2009). Surveys are the most widely used research design. Most of us have certainly encountered surveys before, such as market surveys, student satisfaction surveys and political polls (Bordens & Abbott, 2011). Data are usually collected using questionnaires, structured interviews or structured observations with the aim of generalising from a sample of a population (Bryman, 2012). Three important factors may influence the results of a survey research:
  - *Sampling.* In most cases of quantitative research, it is impractical to survey the entire population. As such, a sample is chosen to represent the population. Choosing the right sample is crucial as it determines the generalisability of research findings or the ability to apply the findings from a sample to a larger population (Bordens & Abbott, 2011). There are essentially two types of sample: probability sample (a sample that has been selected using random selection so that each unit in the population has the same chance of being selected) and non-probability sample (a sample that is not selected using a random selection method, implying that some units in the population are more likely to be selected than others) (Bryman, 2012). Ways to generate probability and non-probability sampling methods are beyond the scope of this text (readers are encouraged to further explore research methods by consulting established literature such as Bryman (2012) and Cooper and Schindler (2008)). As each type of sample may generate different results, researchers should explain the sampling process and its limitations in their reports.
  - *Validity of instrument.* Validity refers to the issue of whether an instrument that is devised to measure a variable really measures that variable (Bryman, 2012; de Vaus, 2001). Common ways to establish validity are (Lucko & Rojas, 2010):
    - Face validity, which is a subjective judgement of a non-statistical nature that seeks the opinion of non-researchers regarding the validity of an instrument. For example, a questionnaire to measure safety climate has face validity if safety managers agree that it has included items that are perceived to influence the safety climate in construction projects.
    - Content validity, which is a non-statistical approach that focuses on determining if the content of an instrument fairly represents reality. For example, a checklist to assess the safe installation of a fall protection mechanism has content validity when it has included items considered to be representative of best practices that are generally accepted in the industry.

- Criterion validity, which is the extent to which the results of an assessment instrument correlate with another, presumably related measure (criterion). Criterion validity is established when the findings of a research study agree with the outcomes of related studies, even though the detailed approaches may differ. For example, the findings from a series of structured interviews identify factors that motivate workers to behave safely in which the factors agree with other studies that investigated the motivational effects of work conditions.
- Construct validity, which refers to whether an instrument or a research effort is measuring what it is supposed to measure according to its stated objectives. For example, an instrument to measure attitudes towards safety has construct validity only if the instrument actually measures those attitudes. In this case, the researchers must justify the process by which the instrument and their conceptual contents were developed. A pilot test is a common way to accomplish construct validity by fine-tuning the instrument before its use in the actual data collection.

  o *Reliability of instrument.* Reliability refers to the consistency of the instrument in measuring the variable, that is, the same 'reading' is generated from the instrument if used repetitively (de Vaus, 2001). Reliability is determined by three factors: stability (the consistency of the instrument over time); internal reliability (typically using Cronbach's alpha coefficient to ensure that all items that measure the same variable cohere and are related to each other); and inter-observer consistency (ensuring a consistency in decisions when more than one 'observer' is involved in using the instrument to collect data that require a great deal of subjective judgement, such as classifying behaviour in structured observation and categorising open questions) (Bryman, 2012).

- *Experiments,* which aim to determine if a specific treatment influences an outcome, which can convincingly identify a cause–effect relationship (Creswell, 2009; Leedy & Ormrod, 2013). An experiment typically involves providing a specific treatment to one group and withholding it from another. The performance of each group in relation to a predetermined set of outcomes is then compared and analysed. In an experimental research design, controlling the confounding variables is important to rule them out as explanations for any effects observed. There are some strategies to control these confounding variables, including (Leedy & Ormrod, 2013):
  o Keeping some things constant so they do not account for any differences observed.
  o Using a control group, that is, a group that receives either no intervention or a neutral intervention, and comparing the performance of the control group to an experimental group that participates in an intervention.

  ◦ Assigning people randomly to groups so that researchers are able to reasonably assume that the groups are quite similar on average and that any differences are due entirely to chance.
  ◦ Using statistical techniques to control confounding variables.

Criticism of quantitative research

Despite its popularity, there are common and recurring criticisms of quantitative research. First, quantitative research fails to distinguish people and social institutions from the natural world, which means it ignores the fact that people interpret the world around them, a capacity that cannot be found among the objects of the natural sciences, such as molecules, cells and materials (Bryman, 2012). Second, many researchers argue that quantitative research is value-free, that is, objective, but no one can be fully detached from any type of research because the researchers themselves influence and shape their research, based on certain assumptions about the world through the accumulated knowledge that they have gained (Grix, 2004). Third, bias may occur in quantitative research as the actual behaviour of respondents may differ from their answers (Bryman, 2012). Fourth, quantitative research is generally considered reliable because it aims to control or eliminate extraneous variables within the internal structure of the study, thus allowing the data to be assessed by standardised testing. However, this may become a serious weakness of quantitative research, especially when the data have been abstracted from their natural context or there have been random events which are assumed not to have happened (Carr, 1994).

Qualitative research

Qualitative research emphasises words and meanings rather than quantification in the collection and analysis of data (Bryman, 2012). Rather than attempting to quantify complex social phenomena, qualitative research develops interpretive narratives from their data in an effort to capture the complexity of those phenomena. Because of the subjective nature of qualitative research, researchers begin with open minds and are ready to immerse themselves in the complexity of the situation and interact with their participants. Data are collected from a small number of participants who might be best to shed light on the phenomenon under investigation. Variables and theories are then drawn from the data, explaining the phenomenon in that particular context, which may not be generalisable (Leedy & Ormrod, 2013).

There are six research designs available to conduct qualitative research (Creswell, 2009):

  • *Ethnography* is the art and science of describing a group or culture (Fetterman, 1998). It is particularly useful in gaining an understanding of the complexities of a particular sociocultural group (Leedy & Ormrod, 2013). Rather than focusing on collecting specific data to enable controlled research as in the case of quantitative research, ethnographic research strives to take in as much of the complexity of an environment as possible to refine future observations. Ethnographic research typically requires longer observation periods (months if not years), mostly to minimise the externally imposed variation caused by having an observer present. The detailed experiences gained during these periods give researchers a rich understanding of complex phenomena that occur within that specific social environment so that explicit and implicit meanings of words, actions and artefacts can be developed (Phelps & Horman, 2010). In ethnography, the researcher may do the following or collect data using the following methods (Bryman, 2012):

- Immersing herself/himself in a social setting for an extended period of time, that is, participant observation;
- Making regular observations of the behaviour of members;
- Listening to and engaging in conversations;
- Interviewing key informants on issues that are not directly observable;
- Collecting documents about the group;
- Developing an understanding of the culture of the group and people's behaviour within that group;
- Writing up a detailed account of that setting.

Baarts (2009) conducted an ethnographic study in the context of construction safety. She worked full-time for 7 months as an apprentice in a gang of 30 construction workers at a Danish construction site. She wrote daily field notes using a diary format and a narrative tone to record what she had observed, been told, and experienced. The notes include descriptions of tasks and work procedures, conversations and jokes, accidents and critical incidents, safety rules and general information, as well as the researcher's feelings of wonder, boredom, fear and doubt. At the end of the fieldwork, she conducted 21 semi-structured interviews with her workmates on issues concerning their experiences of risk and workplace accidents. Her analysis of the field notes and interview transcripts found that individualist and collectivist preferences (collective individualism) influence the amount of risk the individual worker will assume and expose workmates to. Self-regulation, self-confidence and independence are acceptable values as far as they do not pose a threat to the solidarity of the community or safety of other workers. She concluded that the informal practice of safety is a tight-rope act that involves balancing the form and scope of collective and individualistic preferences, including the definition of what is too individualistic.

- *Grounded theory* is a systematic development of theory from data through inductive and deductive thinking (Phelps & Horman, 2010). Grounded theory aims to develop a general, abstract theory of a social phenomenon grounded in the views of participants. Typically, it involves constant comparisons of data with emerging theories and theoretical sampling of different groups to find similarities and differences (Creswell, 2009). The major purpose of a grounded theory study is to begin with the data and use it to develop a theory. The term grounded actually refers to the idea that the theory is derived from and rooted in the data collected in the field rather than taken from the review of literature. Data can be collected using a variety of methods including interviews, observations, documents, historical records and videos (Leedy & Ormrod, 2013).

Choudhry and Fang (2008) conducted seven semi-structured interviews in Hong Kong with workers who had been accident victims. A grounded theory approach was adopted to identify emerging themes during the analysis of interview transcripts. They found that workers involved in unsafe behaviour because of a lack of safety awareness, to be one of the 'tough guys', work pressure, co-workers' attitudes and other organisational, economic, and psychological factors. They further argued the significant role of management, safety procedures, psychological and economic factors, self-esteem, experience, performance pressure, job security and education, and safety orientation and training in influencing worker's safety behaviour.

- *Case study* is an examination of a single individual, family, organisation, event, activity, or process in depth for a defined period of time (Rubin & Babbie, 2011). A study of a single case may seem to have limited applicability, but its unique or exceptional qualities can promote understanding or inform practice for similar situations. A variety of data collection methods, including questionnaires, interviews, observations and data mining of documents, can be employed to gain in-depth and detailed understanding concerning the case under investigation (Leedy & Ormrod, 2013).

  For example, Zhou et al. (2011) carried out two questionnaire surveys, 3 years apart, in a Chinese construction company to identify effective factors that generate safety climate improvement. More than 600 workers participated in each survey. Analysis results of both surveys were consistent, in which a four-factor structure of safety climate was identified. Significant improvement was also found on the four identified factors over the three-year period. Interviews with safety management officers of the company revealed that factors that stimulated the improvement were the enforcement of vigorous safety rules and regulations and the increased intensity of safety training and safety promotion.

- *Phenomenology* is a research design which aims to understand people's perceptions, perspectives and understanding of a particular situation (Leedy & Ormrod, 2013). It is a mode of seeing a phenomenon that uses both intellectual and emotional sensibilities with the aim to gain a more whole and comprehensive understanding (Seamon, 2000). A lengthy, unstructured interview with people who have had direct experience with the phenomenon being studied is a typical method adopted in a phenomenology study. During an interview, it is essential for the researcher to be alert for subtle yet meaningful cues in participants' expressions, pauses, questions and occasional sidetracks (Leedy & Ormrod, 2013). The use of phenomenology in construction engineering and management research is rare, although phenomenology has been used to understand the person–environment relationship in the design-related discipline (Seamon, 2000).

- *Narrative* is a study of the lives of individuals. The researcher asks one or more individuals to provide stories about their lives and then the researcher often retells the stories into a narrative chronology which combines views from the participants and the researcher (Creswell, 2009). In construction safety, a narrative may be used to describe an accident based on eyewitnesses' testimonies. The researcher then presents the narrative in a logical manner, which also includes the root causes of the accident and the recommended solutions.

- *Content analysis* is a detailed and systematic examination of the contents of a particular body of material, for example, books, newspapers, journals, articles, legal documents, Internet blogs, films, art and music, for the purpose of identifying patterns, themes or biases. Typical steps in a content analysis are:
  1. Identifying the specific body of material to be studied. A sample may need to be selected when this body is large.
  2. Defining the qualities to be examined in precise, concrete terms.
  3. Breaking down each item into small, manageable segments that can be analysed separately when the material involves complex or lengthy items.
  4. Scrutinising the material, for examples of each quality defined in step 2. More than one reviewer may be needed when judgements are more subjective in nature.
  5. Tabulating the frequency of each quality found in the material being studied. Thereafter, statistical analyses can be performed to interpret the data and link the results to the research problem. This step implies that a content analysis essentially has both quantitative and qualitative research characteristics.

Criticism of qualitative research

As with quantitative research, there are also some criticisms of qualitative research. First, many researchers consider the use of limited samples to build an argument as a weakness, particularly concerning the representativeness and generalisability of the research. Second, critics argue that qualitative research lacks objectivity and has a tendency to use personal opinions instead of evidence to support arguments (Grix, 2004). Third, some critics also point out that qualitative studies are difficult to replicate because they are unstructured and often reliant upon the researcher's ingenuity (Bryman, 2012).

Table 7.1 presents common distinctions between quantitative research and qualitative research, developed on the basis of several research method texts including Bergman (2008), Cooper and Schindler (2008), Creswell (2009), and Grix (2004). Quantitative and qualitative research principles give rise to different approaches with different emphases. Quantitative research relies on objectivity and emphasises on precisely measuring variables and testing hypotheses, whereas qualitative research is interpretive and emphasises the detailed examination of specific cases that arise in the natural flow of social life, sometimes also to generate new hypotheses. These differing assumptions and emphases influence the characteristics of quantitative and qualitative data. Soft data, for example, words, sentences, pictures and symbols, dictate qualitative research, while quantitative research collects hard data in the form of numbers. In short, quantitative research and qualitative research are different in many fundamental ways, including logic, research path, mode of verification and the way to arrive at a research question (Neuman, 2011).

Mixed methods research

Mixed methods research is a type of research that integrates quantitative and qualitative methodologies within a single research design. This need not necessarily refer to the combination of research methods associated with one research methodology, but could involve the combination of methods that transcend different methodologies (Bryman, 2012). Many researchers believe that both methodologies complement rather than rival each other, and thus, qualitative research can compensate for the weaknesses of quantitative research and vice versa (Cooper & Schindler, 2008; Neuman, 2011). There are three approaches to mixed methods research (Bryman, 2012):

- *Triangulation.* The use of quantitative research to corroborate qualitative research findings or vice versa.
- *Facilitation.* One research methodology is employed to aid research using the other research methodology.
- *Complementary.* Two research methodologies are employed so that different aspects of an investigation can be merged.

There are three arguments for the use of mixed methods research. First, the high complexity and multi-dimensionality of real-world situations require the use of multi-methodology to be investigated and dealt with effectively. Second, a phenomenon under investigation typically does not consist of a single event but a process that proceeds through stages. One methodology could be more useful in relation to some stages than the other; thus combining methodologies may produce a better result. Third, many people have actually combined methodologies in practice; therefore it is necessary to further consider the philosophical and theoretical aspects of multi-methodology (Mingers & Brocklesby, 1997).

Table 7.1 Distinctions between quantitative research and qualitative research

Feature	Quantitative research	Qualitative research
Relationship between the researcher and research participant	The possibility and necessity of separating the researcher from the research participant	An interdependence between the researcher and research participant
Research focus	Finding out numerical qualities of an event or case	Understanding the nature and essence of an event, person, or case
Research purpose	- Predict, describe, test theory - Tackle macro-issues, using large, random, and representative samples - Identify general patterns and relationships	- Understanding and theory-building - Tend to analyse micro-issues, using small, non-random, and non-representative samples - Interpreting events of significance
Research design	- Deductive - Surveys and experiments	- Inductive - Ethnography, phenomenology, grounded theory, case study and narrative
Research methods	- Questionnaires (close-ended questions) - Structured interviews or observations	- Open-ended questions - In-depth interviews - Participant observation
Samples	Tend to be large, representative samples	Tend to be small, non-representative samples
Analysis and finding	- Computerised analysis dominated by statistical and mathematical methods - Clear distinction between facts and judgments - Findings rely heavily on the quality of the data collection instrument - Findings attempt to be comprehensive, holistic and generalised	Human analysis following computer or human coding Tend to consider the contextual framework, which makes distinction between facts and judgments less clear Findings depend on how the researcher can probe deeper during data collection Findings are seen to be deep, precise, narrow and not generalised

Criticism of mixed methods research

Although mixed methods research may appear to offer a solution to the deficiencies of individual research paradigms, it is also a subject of criticism. Critics argue that methods generate unique types of findings and knowledge which may not be merged. Some also say that quantitative and qualitative methods are rooted in separate paradigms and so could be considered as incompatible. Despite these criticisms, it should be noted that the notion of research methods carrying fixed philosophical assumptions is difficult to sustain because each method could be used in a wide variety of tasks in both qualitative and quantitative research (Bryman, 2012). Another issue of mixed methods research is that of contradictory findings between the quantitative and qualitative analyses. Strategies that can be employed to solve this issue are collecting additional data, re-analysing original data, giving priority to one form of data and using the results to recommend future studies (Creswell et al., 2008).

Current state of play on safety research methodologies

A meta-analysis by Zou et al. (2014) on the 88 construction safety research papers published in 2009 showed that quantitative research is the prevalent methodology adopted. This also reflects the situation in social research as claimed by Bryman (2012). Similar results were also obtained by Dainty (2008) with regard to construction management research in general. The popularity of quantitative research implies that organisational factors, such as safety management systems, policy, tools and procedures, are the main objects of construction safety research. This philosophical standpoint sees organisational factors as having to be improved and adhered to in order to maintain and improve safety performance. Quantitative research can also be applied to investigate the influence of human factors on safety. Such research may aim to identify certain characters or competencies that need to be possessed or developed for effectively managing construction safety. It tends to use findings as the basis of generalisation, and may therefore disregard the context in which the findings can be applied. A high percentage of quantitative research may also reflect that past research has focused more on 'what' has happened rather than 'why' and 'how' construction safety problems occurred. It may also show that quantitative research methodology might have been more fundable and implementable, particularly due to its generalisability, as compared to qualitative research studies (Zou et al., 2014).

We would suggest that researchers should recognise the importance of the social and cultural factors in safety learning and safety practices. Specific workplace traditions have a significant role in knowledge and skill development (Baarts, 2009). In this context, researchers need to move one step back and do more fundamental work by exploring how knowledge is constructed in the first place (Tsoukas & Mylonopoulos, 2004). Therefore, it may be prudent for researchers to adopt a more qualitative approach to gain deeper and richer understanding of this process.

Table 7.2 summarises the research methodologies adopted in construction safety research published in the proceedings of CIB W099 conferences in 2009 and 2014. CIB is the International Council for Research and Innovation in Building and Construction and W099 is the Working Commission on safety and health in construction (http://www.cibworld.nl/site/commissions/index .html). CIB W099 is seen as the most influential commission representing researchers around the world and papers published in CIB W099 annual conferences are regarded as reflecting and representing the current state of play of research in safety and health in construction. The CIB W099 annual conferences serve as a platform for researchers to present their research findings which are aimed to advance safety performance in the construction and engineering industry.

As shown in Table 7.2, quantitative research was the dominant methodology in both conferences, although it also indicates that there was a shift from purely quantitative to mixed methods research. Overall, there is no major change in research methodology adaptation between the two conferences, separated by 5 years. 'Review or conceptual' in this context refers to research papers that use literature review or previous studies to formulate theories or develop conceptual frameworks. The large number of reviews or conceptual papers are a normal occurrence because researchers typically use conferences to present their preliminary research effort.

Although a significant amount of qualitative research has been conducted in the field of construction safety, closer observation found that the majority of construction safety researchers used interviews as their main qualitative research methods. A view has been expressed that researchers have become obsessed with interviews as a means of discovery, without considering their limitations (Hammersley, 2003). The issue with interviews is that people may say what should have happened based on their attitudes and beliefs rather than what actually did happen (Leedy & Ormrod, 2013), because of SDB, which will be discussed in the next section. Critics further argue that responses in interviews are heavily influenced by the activities of the interviewer, and so interview participants are more focused on presenting themselves in a positive light, rather than presenting facts about themselves or the social entities under investigation (Hammersley, 2003). This shows the importance of data from different resources to triangulate the inferences and outcomes of these

Table 7.2 Research methodologies adopted in construction safety research in 2009 and 2014

Conference	Total	Quantitative		Qualitative		Mixed		Review or conceptual	
		No.	%	No.	%	No.	%	No.	%
CIB W099 2009	60	22	36.6	15	25.0	4	6.7	19	31.7
CIB W099 2014	60	18	30.0	13	21.7	9	15	20	33.3

interview data (Dainty, 2008) because interviews alone may not reveal the true nature of construction safety issues. Furthermore, the limitations of the qualitative methodology as discussed in previous sections may also expose the need to use the mixed methods research.

In such cases, mixed methods research may offer a way of bringing research and practice closer together rather than the increased use of individual research approaches and paradigms. However, mixed methods research only accounts for around 9% of the construction safety studies reviewed in 2009 (Zou et al., 2014). Although the number has increased to 15% in 2014, this is still very low, given the social nature of construction safety problems and the need to co-produce knowledge with the practice community and to ensure the dissemination of generalisable interventions to projects and organisations. Looking at an object of research from multiple points of view, that is, by employing mixed methods research, has the potential to improve accuracy and stimulate further questioning of existing understanding (Neuman, 2011). This is the case for the current state of construction safety research where knowledge needs to be furthered and new, radical approaches need to be applied to improve construction safety performance. Mixed methods research design provides an alternative to mono-method research and may enrich, and generate more reliable, research results (Bergman, 2008; Mingers, 2001).

Social desirability bias in research design

As discussed in the previous sections, safety research can be classified as social research; therefore it is important to consider the possible effect of SDB in research design. However, our review of the current research in safety in construction and engineering shows very few studies have considered this issue in their research design. In this section, we provide a brief discussion on SDB and techniques for minimising SDB in such research.

Social desirability bias, or SDB, is defined as the general tendency to present oneself in a socially or culturally desirable manner that follows the accepted standards of behaviour (Nederhof, 1985). In other words, SDB is the tendency of people trying to make themselves appealing, regardless of their real behaviour or perceptions. SDB happens because people want to be socially acceptable and favourable. As such, SDB may jeopardise the authenticity and reliability of self-reporting surveys and other non-experimental research with misleading answers (van de Mortel, 2008). SDB was first identified and taken into consideration in the 1950s (Edwards, 1953). Since then, it has been discussed and applied in various research fields, including general social science (Bernardi, 2006), medical research (Davis et al., 2010; Leite & Beretvas, 2005), business research (Chung & Monroe, 2003; Dunn & Shome, 2009) and safety science (Sullman & Taylor, 2010).

Construction safety research is prone to SDB and it is likely to be a major issue in construction safety research due to its inherent nature. The severity of

SDB in construction safety research is due to the methodology used in this field. Generally there are two strings of methodology in construction safety research fields; one is experimental and the other non-experimental (Hogg & Vaughan, 2011). Experimental research can be taken both in the lab and on the field, under carefully designed procedures to obtain objective data, usually from equipment or devices, while non-experimental research involves the interactions between examiners and respondents by means of delicately designed surveys, questionnaires, case studies, and so on, to obtain subjective statistics given out by respondents. The prevailing research methodologies in this specific research field include questionnaires and surveys, which are both subjective, self-reporting methods, instead of experimental and modelling methods, as in natural science or experimental psychology. Non-experimental methodologies have become popular for a reason. First, experiments are difficult to conduct due to the complexity of on-site environments while self-reporting questionnaires and surveys are much more accessible and easier to organise; secondly, research in construction safety management requires a good deal of attitudinal measurement, which is difficult to obtain by experiments.

While researchers have acknowledged this situation, there is little evidence to show that effort has been taken to examine and minimise SDB in questionnaire design and implementation, and also data analysis. The possibility of SDB in construction safety research is due to its social sensitivity and the research methodology adopted, such as non-experimental self-report surveys and questionnaires (Hogg & Vaughan, 2011).

As stated earlier, we found that quantitative methodology is the main research methodology in conducting construction safety research. Further analysis revealed that survey questionnaires were the main methods used for data collection in quantitative research, and more than half of qualitative research used interviews as the main qualitative research method (Zou et al., 2014). The complexity of the construction environment makes it difficult to conduct experiments. Although self-reporting questionnaires and surveys are convenient and inexpensive, respondents may give out false answers intentionally or unintentionally and the results may be contaminated and therefore not useful (Hogg & Vaughan, 2011). Furthermore, attitudinal measurement in construction safety research is difficult to obtain by experiments, and the answers provided through questionnaires may be politically correct answers rather than representing the individual's true intentions (Crowne & Marlowe, 1964). In addition, there are ethical, moral and legally sensitive topics in construction management, which are prone to SDB (Roxas & Lindsay, 2011), especially with two kinds of questions – one is related to attitude and perceptions and the other is related to past sensitive behaviours. Despite the importance of considering SDB in research design, it appears that SDB has been largely neglected in construction safety research and papers taking SDB into consideration are limited. Only a small number of researchers appear to have considered SDB. Among these is a study about the perceptions of personal vulnerability to workplace hazards (Caponecchia & Sheils, 2011), which analysed construction workers' optimism bias and speculated that social desirability might be a difficulty in measuring safe or precautionary behaviour together with a measurement on environmental protection behaviours (Chao & Lam, 2011), which tested the effects of SDB on its measurements and suggested the risk of self-reported responsible environmental behaviour. Although SDB research in fields of psychology or social science methodology provides valuable improvements on scale validation, to validate the original questionnaire and to discard contaminated cases while confirming the sustainability of the rest, in construction safety research, it is important not only to identify whether statistical data obtained from questionnaires and surveys are contaminated by SDB, but also to control SDB and minimize its effects to secure the reliability of non-experimental research methods.

Why and how social desirability bias happens

Since SDB was first identified and taken into consideration in the 1950s (Edwards, 1953, 1957), it has been discussed and applied in various research fields. In general social science, the culture influence and gender variance on the strength of SDB were discussed (Bernardi, 2006). SDB was reported in medical research, for example, respondents were found over-reporting their physical activities when compared to data automatically obtained from devices (Adams et al., 2005; Motl et al., 2005); similarly, in dental care research, it was found that the answers in a questionnaire concerning how well respondents protected their teeth were different when compared to dental insurance records (Leite & Beretvas, 2005); dietary intake, in which the SDB was examined and confirmed by two kinds of recall questionnaires (Hebert et al., 1995); alcohol consumption, in which a 20–30% less consumption and 50% probability of failure to report risky drinking behaviour in college students (Davis et al., 2010); and clinical patient treatment such as pain report and depression, in which the implications of SDB on self-reported psychological distress among chronic pain patients and this influence should be concerned (Deshields et al., 1996; Logan et al., 2008). Business research also evaluated the effects of SDB on accountant ethics, and discovered that ethical situations are sensitive to SDB (Chung & Monroe, 2003), cross(Dunn & Shome, 2009). The SDB related research in safety science appeared in driving behaviour. For example, undergraduate students were asked to fill questionnaires about their driving behaviours publicly and privately and it was found that SDB existed in self-reported data (Sullman & Taylor, 2010).

Why social desirability bias happens

To explain why SDB happens, two questions should be answered: (a) why people want to be socially acceptable; and (b) why people give false answers.

For the first question, human beings are social individuals: externally, they live in groups or circles; internally, they have self-knowledge and self-esteem (Myers, 2010). The social normative influence and the desire to be liked makes people obey socially accepted standards and avoid rejection (Hogg & Vaughan, 2011). Being rejected by the group or society can be painful. Brain scans showed that group judgments could activate the same brain area as the one activated by the pain of bad betting decisions (Klucharev et al., 2009). Social psychology also points out that conformity is greater when people respond publicly (Hogg & Vaughan, 2011). As a result, confidentiality protection becomes important in eliminating such pressure.

In the perspective of self, self-esteem pushes people to present their best self. It gives people portraits of themselves and motivates them to pursue the ideal image they want to be (Myers, 2010). However, the side effect of self-enhancing is that it can lead to self-serving bias, the tendency to perceive themselves as favourable. It can also lead to 'self-presentation', the desire to present a socially favoured image to the people around them as well as themselves (Myers, 2010). In short, both social influence and self-enhancing are associated with SDB in various ways.

The theory of cognitive dissonance by Festinger et al. (1956) can explain why people give out fake answers (Näher & Krumpal, 2011). Cognitive dissonance is the discomfort of conflicting perceptions at the same time, due to various information sources and experience. There are three mechanisms to reduce dissonance according to Festinger (1957): new cognitions can be added, such as excuses; some cognition can be subtracted, that is, forgotten or ignored; some cognition can be substituted, for example, the negative impact can be replaced by the positive impact (Hogg & Vaughan, 2011). In answering questionnaire surveys, when respondents' attitude, perceptions and behaviour differ from the established social norms or image of them, discomfort appears; therefore, they tend to relieve the dissonance by changing their cognition (Festinger, 1957). In other words, they tend to cheat and give out false answers that abide by the social norms, ethics, regulations, laws and non-codified norms within their groups.

How social desirability bias happens

SDB happens in socially sensitive situations. If respondents feel secure to answer insensitive questions, they are much less prone to cognitive dissonance. From the analysis on the causes of SDB, it can be inferred that there are two situations. One is the situation when the respondent unconsciously replaces the truth with a desirable fake answer due to the self-serving bias. The other is due to the need of self-presentation that makes people consciously present themselves in a desirable manner to avoid social rejection. Figure 7.2 provides a summary of the points discussed above.

In research on different categories of SDB (Paulhus, 1984), the author distinguished self-deceptive enhancement (SDE) from impression management (IM). SDE refers to the situation in which respondents unintentionally reply with a fake answer because they actually believe their responses are real, which

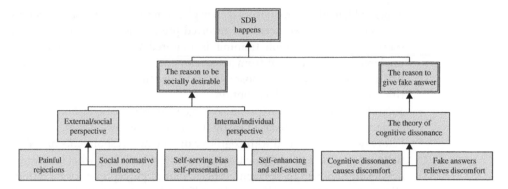

Figure 7.2   Why and how social desirability bias happens

associates with self-serving bias. IM refers to the situation in which respondents intentionally reply with false answers because they want to build up favourable figures by self-presentation (Paulhus, 1984). While IM can be detected and controlled, bias from SDE is only detectable but unavoidable (Nederhof, 1985); therefore, strategies should be developed separately. It should be noted that SDB is difficult to exclude completely from self-reporting techniques.

The majority of construction safety research uses non-experimental methods. As convenient and inexpensive as they are, non-experimental methods have their inherent side effects compared to experimental methods. While in experiments, data are collected through devices and are thus objective, in non-experimental social science researches where subjective self-reporting is the main technique to gather data, structured questions and limited choices are, as we have already stated, prone to SDB. In questionnaires, respondents can easily fake their answers by selecting wrong choices as respondents are not supposed to be interrupted during the process of completing the questionnaire; worse still, as data are analysed after forms are collected, and SDB is identified, the results are contaminated and may not be used (Hogg & Vaughan, 2011). In surveys, respondents may also fake their answers by lying to researchers; even if the researchers speculate that respondents lied, it is disallowed to change their answers, and data collected may again not be usable.

Additionally, there are quite a few socially sensitive topics in construction safety research that are prone to SDB. The more ethical the question is, the more SDB can be expected (Chung and Monroe, 2003). Moral or legal topics can generate SDB as well (Roxas and Lindsay, 2011). Two kinds of questions are strongly prone to SDB:

- *Safety attitude and risk perceptions.* For example, when asked whether labourers are willing to wear personal protective equipment or managers are willing to implement safety management systems, it is highly unlikely to get negative answers because respondents are aware of what is favoured by society, the law and regulations and ethics. Similarly, answers relating to job satisfaction may be skewed by SDB because respondents are afraid of being blamed by supervisors if they give a negative response. Questions of whether on-site workers have received pre-job safety training is legally sensitive because pre-job training is required by the law, though the implementation rate is far from satisfactory; as a result, it is quite possible that respondents would choose 'yes' to show that they have obeyed the rules, to avoid perceived retribution.

- *Subjective opinions or perceptions.* A large proportion of questions may be subjective which can be faked easily (Crowne & Marlowe, 1964). As stated before, when confronted with cognitive dissonance, respondents may fake their answers so as to mitigate the discomfort of betraying what is believed or socially accepted to be favoured behaviour or perceptions, or intentionally try to build a better self-image even if they have no personal relationship with the researchers.

Group culture may aggravate the situation, especially in eastern cultures where collectivism is highly valued. As is illustrated by several researchers, culture is an important factor of SDB, and collectivism is positively related to the degree of SDB (Bernardi, 2006). As a result of the pressure of collectivism posed on individuals, workers are more prone to worry about their image in the social context and suffer from greater pain from the possibility of social rejection; also, the sense of group in the collectivism culture trains people to defend their group, self-servingly believing consciously or unconsciously that their group or organisation behaves in a socially desirable manner; therefore, they are more likely to hide the true answer and respond in a way that is more acceptable socially because they do not want to compromise the image of themselves and also the public reputation of the organisation as a whole.

Techniques for minimising social desirability bias in safety research

There are several techniques to minimise SDB. First, indirect questions can attenuate SDB by asking respondents what other people think (Jo et al., 1997). Secondly, 'forgiving wording' decreases SDB by giving out excuses for cognitive dissonance (Näher & Krumpal, 2011). For example, the 'everybody approach' type of statement, which indicates that a specific situation is the norm so that respondents feel forgiven (Barton, 1958). Similarly, a permissive context provides a context in which sensitive answers are permissive socially so that respondents feel free to answer honestly. Thirdly, a mathematical approach called randomized response decreases SDB (Lensvelt-Mulders et al., 2005). The researchers can infer the real response from deliberately false data, but it can only be used at a group level (Caponecchia & Sheils, 2011). More privacy protection methods can also be used, including the numbered card, with the cards being answer tokens; the sealed ballot technique, with a sealed box to

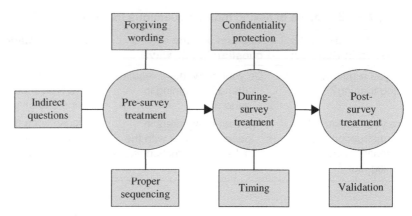

Figure 7.3 Techniques for minimising social desirability bias (SDB) in safety research

put answers in (Barton, 1958); and the informal confidential voting interview (Gregson et al., 2002).

Based on the discussions presented above and in the previous sections, a three-stage (pre-, during- and post-survey stages) SDB minimisation technique for construction safety research is proposed, as shown in Figure 7.3. The techniques follow two principles: one is to relieve cognitive dissonance and the other is to help build a confidential and secure environment. The approaches to relieve dissonance include (1) adding new cognition by providing excuses; (2) substituting cognition to relieve dissonance; and (3) relieving the social context by creating a confidential environment. The focus of pre-survey control strategies is to design the questionnaire delicately to minimise SDB; the focus of survey control strategy is to make respondents feel less uncomfortable and more confident to be honest; and the focus of post-survey control strategy is to examine the validation of the focal questionnaire.

Technique 1 – asking indirect questions

There are two kinds of indirect questions. One is to ask about the other persons' opinion instead of the respondent's own opinion. This technique delicately substitutes the perception of respondents themselves and reduces dissonance. Due to the effect of 'projection', indirect questions are able to probe the respondents' actual perception without their awareness. The other is to ask the respondents a set scenario instead of a direct general question. Two examples are given in Table 7.3.

Technique 2 – using forgiving wording

Forgiving wording provides a context in which unwelcome answers are acceptable, so respondents may feel more comfortable to give such answer. There are

Table 7.3   Examples of asking indirect questions to minimise SDB

Instead of asking the question in this way:	You can ask the question in this way:	Changes made:
Do you regularly attend job training?	Do your <u>co-workers</u> regularly attend job training in general?	Asking what they feel about 'a typical other' feels, instead of 'you'
Are you familiar with construction site risk identification and mitigation?	Are your <u>co-workers</u> in general familiar with the risks in construction sites?	Asking what they feel about 'a typical other' feels, instead of 'you'

Table 7.4   Examples of forgiving wording to minimise SDB

Instead of asking question in this way:	You can ask question in this way:	Changes made:
Are you used to wearing PPE?	Many people are not used to wearing PPE. How about you?	Providing a context that many people are doing it
What is your view of an accident?	Accidents happen from time to time in construction projects. What is your view of an accident?	Providing excuses

two ways of doing so; one is to provide excuses, especially irresistible ones, and it could relieve the internal pressure; the other is to provide a context that everyone else behaves in an unwelcome way and it could relieve the external pressure. Two examples are given in Table 7.4.

Technique 3 – using proper sequencing

Sequencing refers to the technique of carefully arranging the sequence of answers so that respondents would not automatically pick up the first or the last choices that are most socially accepted. Sequencing choices mixes positive and negative statements to avoid doing a simple 'ticking exercise' rather than paying attention to reading and understanding the questions.

Technique 4 – providing confidentiality protection

Social psychology points out that conformity to social pressure is greater when people respond publicly (Hogg & Vaughan, 2011). As a result, SDB is more likely to happen when respondents believe that they are placed in front of an audience. Confidentiality protection eliminates the existence of an audience and becomes important in alleviating social pressure. Confidentiality protection in the pre-survey stage refers to stating clearly at the beginning of the questionnaire that all information obtained from this survey would be sealed and protected as confidential. During the survey, it includes techniques such as explaining directly and clearly to respondents, keeping supervisors away, and so on. It is then unnecessary to worry about punishment from supervisors or group members, and increases the possibility of responding honestly. In addition, appropriate confidentiality protection is a legal and ethical requirement.

Technique 5 – considering suitable timing

Timing refers to giving respondents a limited time to answer each question, so that they present the spontaneous and genuine response, or the first instant 'gut' feelings without second thoughts. It makes the perception process of cognitive dissonance difficult, and respondents are under pressure to follow the cognitive settings provided by the researcher. As a result, the answer is more likely to be true and hence SDB is minimised. This method could be particularly useful if the survey questions are administered through an online platform where questions could appear on the computer screen one by one with a limited time to answer, for example, 3–5 seconds.

Technique 6 – validating social desirability bias

A premise for minimising SDB is the existence of SDB in the responses. Therefore, it is necessary to validate the questionnaire. As discussed in the previous section, a validation scale can be added to the focal questionnaire to examine if the survey data/results are contaminated. Available scales include Marlowe Crowne Social Desirability Scale (MCSDS), the short forms of MCSDS, and the Balanced Inventory of Desirable Responding (BIRD) (Paulhus, 2002). The scores on the SDB scale indicate the degree of SDB of the respondents, and the correlation between SDB scale and the focal questionnaire indicates the extent of influence of SDB. If the correlation between the SDB scale statistics and focal questionnaire data is significant, the data is believed to be biased and not suitable for further analysis. Conversely, if the correlation is not significant, it is safe to use the data for further analysis.

Research-practice nexus

Safety learning and safety practice improvement take place in the social world, among and through other people (Gherardi & Nicolini, 2002; Wadick, 2006). The integration of the realms of theory and practice is needed to ensure that research findings are relevant to the promotion of continual safety improvement. In order to address this issue, a new mixed methods research design for construction safety research is proposed. This is illustrated in Figure 7.4. First,

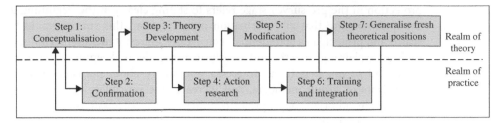

Figure 7.4   Cyclic iterations and nexus between research, theory and practice

recognising the nature of safety learning and practice, the proposed research design incorporates the use of qualitative methods to capture the richness and depth of safety in practical settings. Second, it places an emphasis on knowledge co-production by researchers and practitioners, and third, it encourages regular communication with industry organisations and professionals to promote the dissemination of research findings into existing practice.

Traditionally, research tends to be designed and implemented in isolation from the influence of other parties, particularly those with commercial interests, in the belief that this ensures the objectivity of research findings. In this paradigm, the industry is seen primarily as a potential source of data. The boundary between the realm of theory and the realm of practice is often enforced to avoid influencing the quality of the data (Robinson & Tansey, 2006). Today, however, there is a push for the two realms to engage with each other, that is, to co-produce. The UK government, for instance, encourages social researchers to be heavily involved in policy-making by determining what works and why, and what policy initiatives are likely to be most effective. Academic faculty are also judged on the basis of research that delivers demonstrable benefits to the economy, society, public policy, culture and quality of life (Martin, 2010). In the USA, social researchers are encouraged to address the 'utilisation crisis' issue by interacting with practitioners throughout the research process to ensure that findings are timely and focus on practice issues (Martin, 2010). Similarly, the Australian government provides funding to support research projects which are a collaboration between higher education institutions and partner organisations, while also addressing the national strategic research priorities (Australian Research Council, 2012).

Step 1: conceptualisation

With reference to Figure 7.4, the research can actually be commenced at any step in the model which essentially depends on the research problem/question, that is, derived by a theoretical or industrial problem. For convenience and ease of understanding, however, we explain the process from step 1, conceptualisation. This step is about examining past studies and theories related to the investigation area which aims to gain a firm grasp and deeper understanding of the rationale and background of the problem under scrutiny. As such, research problems need to be clearly identified and research objectives must be formulated to solve the problems. A conceptual or theoretical model may need to be generated to give an overview of the research and show the relationship between various independent and dependent variables that will be investigated.

Step 2: confirmation

The next step is confirmation, to make sure that the research problems and objectives are relevant and in accordance with the need of construction industry/practitioners. The aim of this step is to comprehensively understand the nature of construction safety practices and the context of the issue that will be investigated. Quantitative, qualitative or mixed methods can be used, depending on the type of data that are needed to address the research problems or answer the research questions. However, to gain a richer understanding of the issue, as well as to reflect the exploratory nature and inductive reasoning of the investigation at this step, qualitative methods offer a route to the deeper, contextualised understanding needed (Green et al., 2010). Although interviews would appear to be the most convenient qualitative method for this purpose, it also has weaknesses as discussed earlier. The inclusion of other qualitative methods may be more effective to capture the real condition of construction safety practices. A targeted ethnographic research design, or at least some form of observation, can help to understand the need of the construction organisations involved in the research and to confirm or amend the conceptual framework developed in the previous step, so that the framework is aligned with the industry needs.

Ethnography can be considered as an art and science of describing a group or culture. The focus of inquiry in ethnography is to seek predictable patterns of human thought and behaviour (Fetterman, 1998). Ethnography places researchers in the midst of whatever it is they study; thus researchers can examine and participate in various phenomena as perceived by participants and represent these as accounts (Berg, 2009; Phelps & Horman, 2010) Ethnography can offer rich and practical understanding concerning the complexity of informality in safety learning processes (Baarts, 2009). Researchers can learn through either sharing or empathising with the experiences of construction professionals and/or workers. By using a theoretical model developed in the conceptual stage, ethnography can provide routes to dig deeper and go beyond what is immediately observable, and practical suggestions regarding workplace interventions can be formulated (Pink et al., 2010).

Observation is another qualitative method that should be considered when ethnography is impractical. One of the main strengths of observation is the ability to collect original data at the time of occurrence. It allows researchers to capture data without depending on reports by others. Furthermore, research participants seem to accept observation as less intrusive than the direct questioning approach, hence it reduces bias in the data (Cooper & Schindler, 2008). Observational studies are able to provide answers to 'what' phenomena occurred and give insights into 'why' the phenomena occurred (Leicht et al., 2010). Observation is particularly useful when people are involved in a process, which is also a key target of construction safety research.

Step 3: theory development

Based on the findings from the confirmation step, step three is the collaboration between researchers and industry partners/professionals to formulate theories or models which inform the improvement of safety performance. The principle of grounded theory may be used to formulate these theories or models. Generating grounded theory, that is, the discovery of theory from data, is a way of arriving at theory suited to its supposed uses, which is in contrast to the generation of theory using logical deduction from a priori assumptions. Grounded theory studies are valuable when existing theories about a phenomenon are insufficient or lacking (Leedy & Ormrod, 2013). Through better use of theory-building methods such as this, the construction research community can provide a needed complement to the current prevailing methods and support improvement in this field (Phelps & Horman, 2010).

Step 4: action research

The theories or models are then put into practice through action research in the participating companies, which is step four in the process. Action research is a research approach which aims at building and/or testing theory within the context of solving an immediate practical problem in a real setting (Azhar et al., 2010). Typically, action research is carried out with a team approach that includes researchers and members of organisations (those who are considered as stakeholders in the research effort). The goal of action research is not only for the sake of the research or testing a theory, but also to create a positive social change (Berg, 2009). Both quantitative and qualitative methods can be used during this stage to gather data and information on the effectiveness of the theory or model as well as getting feedback for further improvement. Action research is very useful for conducting applied research in construction and can help improve collaboration between researchers and practitioners, in research and development projects (Azhar et al., 2010).

Step 5: modification

The findings of the action research should result in theory or model modification and improvement. The modified theory or model should be confirmed or verified by practice. As in step three, grounded theory principles can be applied in this step.

Step 6: training and integration

The revised and final theory or model can then be integrated into the participating organisations' practice through communication, intervention and training (step six). Particularly, it has to be integrated into policy strategy and strategic planning to ensure its maximum effectiveness. Furthermore, as discussed in Chapter 5 concerning the nature of safety learning, this communication, intervention and training must be thought through. A participative training setting may be more appropriate and feedback from the participants involved should always be encouraged.

Step 7 generalise fresh theoretical positions

After this, continual review of the theory and model (or framework) is essential to find opportunities for further improvement and to generalise fresh theoretical positions. When a new problem with a need for further research is found, the cycle begins again.

Discussions

It should be noted that this proposed mixed methods research design promotes the application of both deductive and inductive reasoning. It starts deductively by finding problems for research and developing a theory or conceptual model that requires confirmation from industry practitioners. Afterwards, it enters the realm of induction by collecting data in practice to develop a grounded theory, framework or model. This theory is tested before implementation, which indicates the use of deductive reasoning again. Furthermore, this proposed mixed methods research design should not be viewed as rigid, neglecting the research question that should be answered. Each research study requires different kinds of data to be collected and analysed, and as such research methods employed in each step may vary from one research study to another. This mixed methods research design, therefore, should be viewed as a guideline or a plan in which each stage serves as a milestone for researchers to reflect on what needs to be done and how each action can positively impact on practices. Used in this way, the framework presented in Figure 7.4 offers a point of departure for researchers seeking to ensure proper integration of research findings into practice and learning.

At a practical level, the greatest challenge of this proposed mixed methods research design is perhaps gaining support from construction industry/ companies to participate fully throughout the research timeframe. Implementing this mixed methods research design can be an elaborate process and involve a considerable amount of time from research participants. Therefore, it is of paramount importance to find a topic that is truly relevant, useful, value-adding and important to the participating organisations, at the same time having research significance and innovation that will lead to development of theory and contribution to knowledge. Time for implementing the proposed mixed method research design is another factor that should be considered carefully because the research could be unmanageable without proper planning and control.

A greater use of mixed methods research design and multi-methodological research approaches may also benefit construction management research more broadly, particularly that which is oriented towards human factors and the social context of management within the construction sector. However, researchers must also remain cognisant of the disadvantages of methodological pluralism and should not take such decisions without a careful consideration of the inevitable paradigmatic tensions that such approaches invoke.

Assessing the relevance of research outcomes in practical application

In order to increase the research–practice nexus, researchers should engage in the self-conscious integration of theory and practice to assess the relevance of research outcomes in practice. First, researchers should identify research elements, both relevant and irrelevant to practice and the missing elements from the research that would be relevant to practice. Second, researchers should consider the communication for practice which focuses on developing a sense of audience. Table 7.5 illustrates a framework to apply these two principles. The framework summarises the research elements and links them to one another, paving the way to reach conclusions as to the application of research outcomes in practice. Furthermore, the framework also helps researchers to be aware of putative problems in theorisation, method, data analysis and synthesis. Such a framework has been applied in the political science discipline as explained by Evans (2010) and here we have adapted it to the context of safety research in construction and engineering.

Conclusions

In this chapter, we have discussed the fundamentals, together with the criticisms, of quantitative and qualitative research, as well as mixed methods research, in the context of construction safety research and practice. We have also discussed the effect of SDB in safety research and techniques to minimise SDB. In order to improve research–practice nexus in safety in construction and engineering, we have advocated the greater use of mixed methods research design to diversify the knowledge generated and to help integrate the realms of theory and practice to facilitate the collaboration between researchers and practitioners in construction safety. By adopting this approach, it is expected that research findings will become more relevant and useful to construction industry and practitioners, while at the same time contributing to the advancement of conceptual understanding and theory development. In short, this mixed methods research design incorporates several different research methods, inductively and deductively, and encourages iteration between the

Table 7.5   A framework for assessing the relevance of research outcomes in practical application (based on Evans, 2010)

Narrative summary	Verifiable indicators of rigour	Means of verification	Critical reflectivity
A contribution of the research to practice	Identifying the measures that show the potential of the research for improving existing safety practices	Verifying the sources of data and tools used to measure safety performance	Reflecting on the relevance of the research outcomes for improving safety practices
Theoretical approach	Making sure that the theory or approach can be verified	Making sure that the theoretical approach and core propositions are justifiable	Reflecting on potential bias in the theory, possible re-conceptualisation of the theory and amendments to the theory to make sound knowledge claims
Methodology	Ascertaining that the methodology allows for the verification of the theory and the methods are tried and trusted	Using appropriate methodology; reliability and validity of measurement tools	Reflecting on inherent bias in the method, the data collection process and safety performance measurement
Data analysis and synthesis	Maintaining the credibility of results and generating sufficient evidence to support the results	Verifying the data using triangulation and counter-factual approaches	Reflecting on the reliability and generalisability of the analysis and results
Self-conscious integration of theory and practice	Identifying research elements that are relevant or irrelevant to safety practices; identifying missing research elements which could have been relevant to safety practices; making sure that research results are communicated and accessible to industry practitioners		

realms of theory and practice; thus it demands the combination of multiple forms of research in order to provide insights into the 'what', 'why' and 'how' types of research questions. It may also be beneficial to consider the adoption of multi-methodological approaches to investigate and gain richer insights that are more likely to resonate beyond the specific context from which they originated. We have also provided a framework for assessing the relevance of research to practice.

It should be noted that mixed methods research is not necessarily superior to mono-method or mono-methodology research. Poorly conducted research will generate dubious findings, regardless of how many research methods are employed. Mixed methods research may also produce contradictory or incompatible findings in which the knowledge generated is incommensurable. As such, researchers must be aware of these challenges and consider strategies to translate the findings into meaningful outcomes. It should also be noted that although this chapter is written within the context of construction safety research and practice, the fundamental research methodologies discussed and the research–practice nexus model proposed should be applicable to other construction management research–practice topics and domains.

References

Adams, S. A., Matthews, C. E., Ebbeling, C. B., Moore, C. G., Cunningham, J. E., Fulton, J., & Hebert, J. R. (2005). The effect of social desirability and social approval on self-reports of physical activity. *American Journal of Epidemiology, 161*(4), 389–398.

Australian Research Council. (2012). Linkage Projects Retrieved 18 October, 2012, from http://www.arc.gov.au/ncgp/lp/lp_default.htm

Azhar, S., Ahmad, I., & Sein, M. K. (2010). Action research as a proactive research method for construction engineering and management. *Journal of Construction Engineering and Management, 136*(1), 87–98.

Baarts, C. (2009). Collective individualism: The informal and emergent dynamics of practising safety in a high-risk work environment. *Construction Management and Economics, 27*(10), 949–957.

Barton, A. H. (1958). Asking the embarrassing question. *Public Opinion, 22*(1), 67–68.

Berg, B. L. (2009). *Qualitative Research Methods for the Social Sciences* (7th ed.). Boston: Allyn & Bacon.

Bergman, M. M. (2008). The straw men of the qualitative-quantitative divide and their influence of mixed methods research. In M. M. Bergman (Ed.), *Advances in Mixed Methods Research: Theories and Applications* (pp. 11–21). Los Angeles: Sage.

Bernardi, R. A. (2006). Associations between Hofstede's cultural constructs and social desirability response bias. [Article]. *Journal of Business Ethics, 65*(1), 43–53. doi: 10.1007/s10551-005-5353-0

Bordens, K. S., & Abbott, B. B. (2011). *A Process Approach: Research Design and Methods* (8th ed.). New York: McGraw-Hill.

Bryman, A. (2012). *Social Research Methods* (4th ed.). Oxford: Oxford University Press.

Burrell, G., & Morgan, G. (1979). *Sociological Paradigms and Organisational Analysis*. London: Heinemann.

Caponecchia, C., & Sheils, I. (2011). Perceptions of personal vulnerability to workplace hazards in the Australian construction industry. *J Safety Res, 42*(4), 253–258. doi: 10.1016/j.jsr.2011.06.006

Carr, L. T. (1994). The strengths and weaknesses of quantitative and qualitative research: what method for nursing? *Journal of Advanced Nursing, 20*(4), 716–721.

Chao, Y.-L., & Lam, S.-P. (2011). Measuring responsible environmental behavior: Self-reported and other-reported measures and their differences in testing a behavioral model. *Environment and Behavior, 43*(1), 53–71. doi: 10.1177/0013916509350849

Choudhry, R., & Fang, D. (2008). Why operatives engage in unsafe work behavior: Investigating factors on construction sites. *Safety Science, 46*(4), 566–584.

Chung, J., & Monroe, G. S. (2003). Exploring social desirability bias. *Journal of Business Ethics, 44*(4), 291–302. doi: 10.1023/a:1023648703356

Cooper, D. R., & Schindler, P. S. (2008). *Business Research Methods* (10th ed.). Boston: McGraw-Hill/Irwin.

Creswell, J. W. (2009). *Research Design: Qualitative, Quantitative, and Mixed Methods Approaches* (3rd ed.). Thousands Oaks, CA: Sage Publications.

Creswell, J. W., Clark, V. L. P., & Garrett, A. L. (2008). Methodological issues in conducting mixed methods research design. In M. M. Bergman (Ed.), *Advances in Mixed Methods Research: Theories and Applications* (pp. 66–83). Los Angeles: Sage.

Crowne, D. P., & Marlowe, D. (1964). *The Approval Motive*. New York: Wiley.

Dainty, A. R. J. (2008). Methodological pluralism in construction management research. In A. Knight & L. Ruddock (Eds.), *Advanced Research Methods in the Built Environment* (pp. 1–13). West Sussex, UK: Blackwell Publishing.

Davis, C. G., Thake, J., & Vilhena, N. (2010). Social desirability biases in self-reported alcohol consumption and harms. *Addictive Behaviors, 35*(4), 302–311.

de Vaus, D. (2001). *Research Design in Social Research*. London: SAGE.

Deshields, T. L., McDonough, E. M., Mannen, R. K., & Millier, L. W. (1996). Psychological and cognitive status before and after heart transplantation. *General Hospital Psychiatry, 18*(Supplement 6), 62–69.

Dunn, P., & Shome, A. (2009). Cultural crossvergence and social desirability bias: Ethical evaluations by Chinese and Canadian business students. *Journal of Business Ethics, 85*(4), 527–543.

Edwards, A. L. (1953). The relationship between the judged desirability of a trait and the probability that the trait will be endorsed. *Journal of Applied Psychology, 37*(2), 90–93.

Edwards, A. L. (1957). *The Social Desirability Variable in Personality and Assessment and Research*. New York: Dryden.

Evans, M. (Ed.). (2010). *New Directions in the Study of Policy Transfer*. Abingdon, UK: Routledge.

Fellows, R., & Liu, A. (2008). *Research Methods for Construction* (3rd ed.). Chichester, UK: Wiley-Blackwell.

Festinger, L. (1957). *A Theory of Cognitive Dissonance*. California: Stanford University Press.

Festinger, L., Riecken, H. W., & Schachter, S. (1956). *When Prophecy Fails: A Social and Psychological Study of a Modern Group that Predicted the Destruction of the World*. Minnesota: University of Minnesota Press.

Fetterman, D. M. (1998). *Ethnography: Step by Step* (2nd ed.). Thousand Oaks: Sage Publications.

Gherardi, S., & Nicolini, D. (2002). Learning the trade: A culture of safety in practice. *Organization, 9*(2), 191–223.

Green, S. D., Kao, C.-C., & Larsen, G. D. (2010). Contextualist research: Iterating between methods while following an empirically grounded approach. *Journal of Construction Engineering and Management, 136*(1), 117–126.

Gregson, S., Zhuwau, T., Ndlovu, J., & Nyamukapa, C. A. (2002). Methods to reduce social desirability bias in sex surveys in low-development settings: Experience in Zimbabwe. *Sexually Transmitted Diseases, 29*(10), 568–575.

Grix, J. (2004). *The Foundation of Research*. Hampshire: Palgrave Macmillan.

Hammersley, M. (2003). Recent radical criticism of interview studies: any implications for the sociology of education? *British Journal of Sociology of Education, 24*(1), 119–126.

Hebert, J. R., Clemow, L., Pbert, L., Ockene, I. S., & Ockene, J. K. (1995). Social desirability bias in dietary self-report may compromise the validity of dietary intake measures. *International Journal of Epidemiology, 24*(2), 389–398.

Hogg, M. A., & Vaughan, G. M. (2011). *Social Psychology* (6th ed.). Frenchs Forest, N.S.W.: Pearson Australia.

Jo, M. S., Nelson, J. E., & Kiecker, P. (1997). A model for controlling social desirability bias by direct and indirect questioning. *Marketing Letters, 8*(4), 429–437.

Klucharev, V., Hytonen, K., Rijpkema, M., Smidts, A., & Fernandez, G. (2009). Reinforcement learning signal predicts social conformity. *Neuron, 61*(1), 140–151.

Leedy, P. D., & Ormrod, J. E. (2013). *Practical Research: Planning and Design* (10th ed.). Boston: Pearson.

Leicht, R. M., Hunter, S. T., Saluja, C., & Messner, J. I. (2010). Implementing observation research methods to study team performance in construction management. *Journal of Construction Engineering and Management, 136*(1), 76–86.

Leite, W. L., & Beretvas, S. N. (2005). Validation of scores on the marlowe-crowne social desirability scale and the balanced inventory of desirable responding. *Educational and Psychological Measurement, 65*(1), 140–154.

Lensvelt-Mulders, G. J. L. M., Hox, J. J., van der Heijden, P. G. M., & Maas, C. J. M. (2005). Meta-analysis of randomized response research: Thirty-five years of validation. *Sociological Methods Research, 33*(3), 319–348.

Logan, D. E., Simons, L. E., Stein, M. J., & Chastain, L. (2008). School impairment in adolescents with chronic pain. *Journal of Pain, 9*(5), 407–416.

Lucko, G., & Rojas, E. M. (2010). Research validation: Challenges and opportunities in the construction design. *Journal of Construction Engineering and Management, 136*(1), 127–135.

Martin, S. (2010). Co-production of social research: Strategies for engaged scholarship. *Public Money & Management, 30*(4), 211–218.

Mingers, J. (2001). Combining IS research methods: towards a pluralist methodology. *Information Systems Research, 12*(3), 240–259.

Mingers, J., & Brocklesby, J. (1997). Multimethodology: Towards a framework for mixing methodologies. *Omega, The International Journal of Management Science, 25*(5), 489–509.

Motl, R. W., McAuley, E., & DiStefano, C. (2005). Is social desirability associated with self-reported physical activity? *Preventive Medicine, 40*(6), 735–739.

Myers, D. G. (2010). *Social Psychology* (10th ed.). New York: McGraw-Hill.

Näher, A. F., & Krumpal, I. (2011). Asking sensitive questions: The impact of forgiving wording and question context on social desirability bias. *Quality and Quantity, 46*(5), 1601–1616.

Nederhof, A. J. (1985). Methods of coping with social desirability bias: A review. *European Journal of Social Psychology, 15*(3), 263–280.

Neuman, W. L. (2011). *Social Research Methods: Qualitative and Quantitative Approaches* (7th ed.). Boston: Pearson.

Paulhus, D. L. (1984). Two-component models of socially desirable responding. *Journal of Personality and Social Psychology, 46*(3), 598–609.

Paulhus, D. L. (2002). Socially desirable responding: The evolution of a construct. In H. I. Braun, D. N. Jackson & D. E. Wiley (Eds.), *The Role of Constructs in Psychological and Educational Measurement*. Mahwah, NJ: Lawrence Erlbaum Associates.

Phelps, A. F., & Horman, M. J. (2010). Ethnographic theory-building research in construction. *Journal of Construction Engineering and Management, 136*(1), 56–65.

Pink, S., Tutt, D., Dainty, A. R. J., & Gibb, A. (2010). Ethnographic methodologies for construction research: Knowing, practice and interventions. *Building Research & Information, 38*(6), 647–659.

Robinson, J., & Tansey, J. (2006). Co-production, emergent properties and strong interactive social research: The Georgia Basin Futures Project. *Science and Public Policy, 33*(2), 151–160.

Roxas, B., & Lindsay, V. (2011). Social desirability bias in survey research on sustainable development in small firms: An exploratory analysis of survey mode effect. *Business Strategy and the Environment, 21*(4), 223–235. doi: 10.1002/bse.730

Rubin, A., & Babbie, E. (2011). *Research Methods for Social Works* (7th ed.). Belmont, CA: Brooks/Cole.

Seamon, D. (2000). A way of seeing people and place: Phenomenology in environment-behavior research. In S. Wapner, J. Demick, T. Yamamoto & H. Minami (Eds.), *Theoretical Perspectives in Environment-Behavior Research* (pp. 157–178). New York: Plenum.

Sullman, M. J. M., & Taylor, J. E. (2010). Social desirability and self-reported driving behaviours: Should we be worried? *Transportation Research Part F-Traffic Psychology and Behaviour, 13*(3), 215–221. doi: 10.1016/j.trf.2010.04.004

Tsoukas, H., & Mylonopoulos, N. (2004). Introduction: Knowledge construction and creation in organizations. *British Journal of Management, 15*(S1), S1–S8.

van de Mortel, T. F. (2008). Faking it: Social desirability response bias in selfreport research. *Australian Journal of Advanced Nursing, 25*(4), 40-48.

Wadick, P. (2006). Learning safety in the building industry. Retrieved from http://www.cfmeu-construction-nsw.com.au/pdf/pwreslearnsafetybldgind.pdf

Zhou, Q., Fang, D., & Mohamed, S. (2011). Safety climate improvement: Case study in a Chinese construction company. *Journal of Construction Engineering and Management, 137*(1), 86–95. doi: 10.1061/(asce)co.1943-7862.0000241

Zou, P. X. W., Sunindijo, R. Y., & Dainty, A. R. J. (2014). A mixed methods research design for bridging the gap between research and practice in construction safety. *Safety Science, 70*, 316–326.

8 Strategic Safety Management

This chapter discusses safety management in construction and engineering from a strategic perspective and approach. It gives a helicopter view of strategic safety management (SSM) and also brings together the different topics which have been discussed in the previous chapters. This chapter covers a variety of topics, including the fundamentals of SSM, the process of designing, implementing and evaluating SSM, and a detailed case study.

Although safety performance in the construction and engineering industry sectors has improved significantly in the past century, recent indicators show that the industry is still facing difficulties in further improving its performance, with injuries and fatalities still happening frequently. Today, major construction and engineering organisations recognise the need to integrate safety into all decision-making processes. We advocate that SSM is a way to achieve the level of integration required, to eliminate and mitigate safety risks and to achieve the desired safety cultural maturity. Safety should be implemented not only for the sake of meeting legal obligations, but as a value-added business amalgamated into overall corporate strategic management.

Hale and Hovden (1998) argued that safety has evolved through three ages: technical, human factors and management systems. Similarly, Hudson (2007) argued that there are three waves of safety development: technology, systems and culture. More recently, Pillay (2014) argued that there are five different ages of safety: technological, behavioural and human factors, socio-technical, cultural, and adaptive or resilience. Lingard and Rowlinson (2005) explained that, although necessary, the development and application of systems, structures, and technology are inadequate to further improve safety performance. People, along with their characteristics, proneness to make mistakes and other human-related factors cannot be entirely separated from the process or the system. In essence,

Strategic Safety Management in Construction and Engineering, First Edition.
Patrick X.W. Zou and Riza Yosia Sunindijo.
© 2015 John Wiley & Sons, Ltd. Published 2015 by John Wiley & Sons, Ltd.

organisations should recognise the need to balance the 'science' and 'art' of safety management.

This chapter is written with the intention of linking the topics discussed in the previous chapters into an SSM framework that enables the integration of the science and art of safety management in the context of construction and engineering businesses and projects.

A strategic safety management framework

There are many schools of thought on strategy. The definition proposed by Johnson et al. (2008) is practical as it emphasises key terms considered important to construction and engineering organisations. They define strategy as 'the direction and scope of an organisation over the long-term, which achieves advantage in a changing environment for the organisation through its configuration of resources and competences with the aim of fulfilling stakeholder expectations'. A strategy has three dimensions that can be recognised in every real-life strategic problem situation (de Wit & Meyer, 2005; Price & Newson, 2003):

- *Strategy context*, which is the set of circumstances under which both the strategy process and content are determined and implemented. It is concerned with the 'where' of strategy, for example, the organisation and environment where the strategy process and content are embedded.
- *Strategy process*, which is the manner in which strategies come about. It is concerned with the 'how', 'who', and 'when' of strategy: how should strategy be made, analysed, formulated, implemented, changed and controlled; who should be involved; and when the necessary activities should take place.
- *Strategy content*, which is the product of a strategy process. It addresses a question such as: 'What should be the strategy for the organisation and its constituent units'?

By applying the above definition and dimensions of a strategy to safety management, we contend that the strategy application context is construction and engineering organisations, construction project management and the industry in general. The strategy process consists of strategy development, strategy implementation and strategy evaluation. The strategy content is the different aspects of safety management that should be integrated into construction and engineering business practices. As mentioned earlier, these aspects are the 'science' and 'art' of safety management. Based on this proposition, we propose an SSM framework as shown in Figure 8.1. The framework was developed by referring to the strategic management model developed in the Construction Management New Directions (3rd Edition) by McGeorge and Zou (2013). The following sections provide details of each of the dimensions of the SSM framework shown in Figure 8.1.

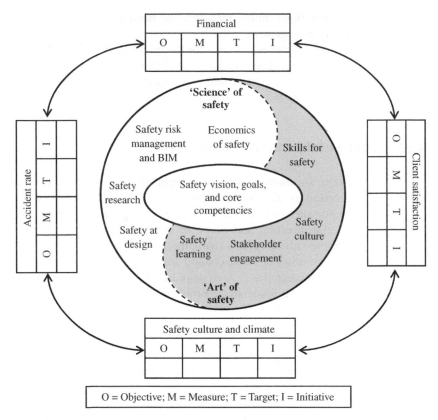

Figure 8.1   Strategic safety management framework

Developing safety management strategies

The first process in SSM is safety strategy development. This involves assessing existing safety strategies, strengths and weaknesses in the organisation, and opportunities and threats in the external environment to formulate new safety strategies, develop implementation plans and determine evaluation methods.

Safety vision, goals, and core competency: principles

As shown in Figure 8.1, the core of the framework includes safety vision, goals and core competencies which can be considered as the foundation and starting point of SSM. In other words, safety should be included in the vision and mission statement as one of the underlying philosophies in organisational operations. Vision is the organisation's ability to see what it will look like at some point in the future. A vision provides a sense of direction and a set of criteria against which actual organisational performance is measured. When it is used properly, a vision can be a powerful motivating force. A classic example is the vision delivered by Martin Luther King Jr. in his 'I Have a Dream' speech. The vision of a better and more just country inspired the American people to take up the cause and make it happen, leading to an unstoppable force of societal change in the USA in the 1960s (King, 1963). Some people, however, misunderstand what a vision is. A vision is not a financial target, but it is broader and achieved over a longer period of time than traditional business goals. A vision is not a one-off, personal effort by the top executives. Although the vision may be developed by top-level managers, it must be shared and embraced by employees in the organisation to become a powerful source for change. A vision is not a solution to a problem because a vision is actually never really attained because visionary leaders continually renew and challenge the boundaries of the vision. Finally, a vision is not merely a catchy phrase, but a fundamental and governing principle for the organisation in its business dealings (Hopkins et al., 2005).

Safety vision, goals, and core competency: case study

In assessing their existing visions, Hamel (2000) suggested organisations ask the following questions: 'Where are we heading to? What is our dream? What kind of difference do we want to make in the world'? Consider Fluor, a FORTUNE top 500 organisation that delivers engineering, procurement, construction, maintenance and project management services around the world. Fluor believes that 'zero incident' is an attainable objective. The organisation employs a range of strategies, which cover management approaches, engagement and training programmes, and safety management system to achieve its safety vision. Although Fluor recognises that the journey is still long, positive results have been manifest along the way (Fluor Corporation, 2014). Another example is Lend Lease and its safety vision, 'Incident and Injury Free (IIF)', which aims to achieve zero incident or injury wherever the organisation has a presence. A case study about Lend Lease's IIF safety management strategy is discussed at the end of this chapter.

As part of the internal analysis, construction and engineering organisations should consider their core competencies and weaknesses in relation to safety. Core competencies are strengths that the organisation has or areas where it does well in comparison with other organisations (Schermerhorn et al., 2014). Core competencies may be found in special safety knowledge or expertise, a robust and established safety management system, a mature safety culture, proper safety equipment and technology, and so on. Weaknesses, on the contrary, are activities that the organisation does not do well or resources it needs but does not possess (Robbins et al., 2012). For example, the organisation may have weak management commitment to safety, limited resources and a bad safety reputation.

In the external analysis, changes and trends in the external environment should be identified. These trends can be positive, which are considered as opportunities, or negative, which are threats (Robbins et al., 2012). Opportunities may manifest in the forms of government regulations that promote safety (although this may be a threat if the organisation does not have the capabilities to meet the new regulations), new safety technology and equipment, safety procedures in the industry, and clients who are more committed to safety. Examples of threats are competition which forces construction organisations to reduce their tender prices and their safety budgets to win jobs, resistance to change among the construction workforce, industry culture and characteristics that hinder safety implementation and poor safety training and learning.

The internal and external analysis together is well known as the SWOT analysis because it is an analysis of the organisation's strengths, weaknesses, opportunities and threats. The next step is to formulate safety strategies. This is about establishing strategy content to capitalise on strengths and external opportunities, protect the organisation from external threats and correct critical weaknesses (Robbins et al., 2012). This strategy content form the components of the 'art' and 'science' of safety management as shown in Figure 8.1 and which have been discussed in the earlier chapters of this book. It should be noted that although the figure indicates a separation between the 'science' and 'art' aspects of safety management, they are interrelated and interdependent in practice. For example, although safety culture is widely accepted as part of the behavioural aspect (art) of safety, a safety management system is a corporate dimension of safety culture which is closely related to the science of safety. Likewise, the application of building information modelling (BIM) is considered as an aspect of the science of improving safety due to its technical nature, but people's attitudes and perceptions on the importance and relevance of BIM will strongly influence its usage and success.

Strategic safety management content

As discussed in the previous chapters, there are many dimensions and aspects of SSM. The following are key elements that should be considered in developing safety strategies:

- *Economics of safety*. Construction and engineering organisations should realise the importance and economic benefits of investing in safety management. Likewise, clients who have the economic power to facilitate safety implementation should also support safety management. Through this attention to safety, designers, contractors and clients will eventually reap its economic benefit.
- *Safety culture*. The concept of safety culture has been widely investigated since its introduction in 1986. Its dimensions and preceding subcultures have been discussed in Chapter 3. There is a tendency, however, to focus only on safety culture within an organisation. This is inadequate because of the nature of the industry, where subcontracting practice and the involvement of numerous stakeholders are common. Therefore, developing safety culture across the supply chain, that is, inter-organisational safety culture, could be the next challenge to deal with (Fang & Wu, 2012). Furthermore, some organisations are operating globally and facing differing cultural backgrounds which will greatly influence the interpretation and

implementation of safety policy and safety systems. More cross-cultural research is needed to achieve the desired integration of safety strategies across business units and cultures.

- *Skills for safety.* The development of safety culture requires the support and leadership of employees, particularly those in safety critical positions, so that safety implementation is aligned from the top to the lowest management level. Zou and Sunindijo (2013) have developed a model which comprises the essential skills for providing safety leadership. This model includes four dimensions, namely, conceptual, human, political and technical skills. It also provides strategies for developing the necessary skills. In their model, the foundational skills are self-awareness, visioning and apparent sincerity. The first-tier mediator skills are scoping and integration, and self-management. The second-tier mediator skills are social awareness, social astuteness, and relationship management.

- *Safety training and learning.* The principles of andragogy should be included in the development of safety training programmes. We should also realise that safety learning happens not only in a formal setting, but also informally via interactions at the workplace and site with people and artefacts at work. Lastly, construction organisations should measure the effectiveness of their training programmes by using the four-level evaluation process as proposed by Kirkpatrick (1979, 1996), which includes recommendations on the training programme and the trainers, the knowledge and skills gained, safety attitude and behavioural changes, and safety culture changes in the longer term.

- *Safety in design and risk management.* There is only so much that can be done in terms of safety during the construction stage. Therefore, safety should be considered during the design stage when, for example, BIM can be used as a technique to detect potential safety hazards. Safety risk management processes and procedures are also useful tools for implementing a safety-in-design concept.

- *Stakeholder and supply chain engagement.* Construction businesses and projects involve many different stakeholders and supply chain members. To ensure the highest safety performance is achieved, it is necessary to engage the key stakeholders and integrate the supply chain members at different stages of projects. For example, it is necessary to engage the clients, designers, contractors and key subcontractors in the design and concept stages to minimise and mitigate safety risks as early as possible. It is also necessary to engage and involve facility management personnel at the early stage of the project so that their safety needs are considered during design and construction.

- *Safety research–practice nexus.* Safety performance should not remain stagnant when improvement is still achievable. Undertaking safety research is a way to promote continuous safety performance improvement. As such, we propose that safety research needs to consider and engage the practice perspectives which we have termed 'research–practice' nexus. Only by having such a nexus can the research outcomes and findings have more 'buy-in' by the practitioners and be more relevant and useful for practice (Zou et al., 2014a; Zou et al., 2014b). The research design supports the use of mixed methods research to generate objective and generalisable research findings while concurrently capturing the richness and depth of safety in practical settings. The research design also places an emphasis on knowledge co-production by researchers and practitioners, promoting the dissemination of research findings into existing practice (Zou et al., 2014b).

- *Safety law and regulations.* Though this book does not have a chapter dealing with this issue, it does not mean it is not important. Relevant laws and regulations are essential to any business operation as we would commonly understand and agree. Managers have the 'duty of care' and all safety risks must be treated at the level of 'as low as reasonably practical'.

Implementing safety management strategies

No matter how well developed strategies are, supporting structures, appropriate procedures, good allocation of tasks and the right people are needed to ensure their successful implementation. Senior managers should enthusiastically support the strategies and communicate those strategies regularly to employees and relevant stakeholders (Schermerhorn et al., 2014). There are three essential factors that support the implementation of strategies:

1. corporate governance
2. organisational structure
3. strategic leadership.

Corporate governance: principles

Corporate governance is 'the framework of rules, relationships, systems and processes within and by which authority is exercised and controlled in corporations' (ASX Corporate Governance Council, 2007, p. 3). It is concerned with identifying ways to ensure that strategic decisions are made effectively, aiming to ensure that the interests of top-level managers are aligned with the interests of shareholders. Therefore, corporate governance reflects an organisation's standards and values (Hitt et al., 2005). The Australian Securities Exchange (ASX Corporate Governance Council, 2007) introduced a Corporate Governance Guideline in August 2007. A summary of the principles set in the guideline are as follows:

- Principle 1 – Lay solid foundations for management and oversight: Companies should establish and disclose the respective roles and responsibilities of the board and management.

- Principle 2 – Structure the board to add value: Companies should have a board of an effective composition, size and commitment to adequately discharge its responsibilities and duties.
- Principle 3 – Promote ethical and responsible decision-making: Companies should actively promote ethical and responsible decision-making.
- Principle 4 – Safeguard integrity in financial reporting: Companies should have a structure to independently verify and safeguard the integrity of their financial reporting.
- Principle 5 – Make timely and balanced disclosure: Companies should promote timely and balanced disclosure of all material matters concerning the company.
- Principle 6 – Respect the rights of shareholders: Companies should respect the rights of shareholders and facilitate the effective exercise of those rights.
- Principle 7 – Recognise and manage risk: Companies should establish a sound system of risk oversight and management and internal control.
- Principle 8 – Remunerate fairly and responsibly: Companies should ensure that the level and composition of remuneration is sufficient and reasonable and that its relationship to performance is clear.

Hitt et al. (2005) explained that there are three internal governance mechanisms and a single external one typically used in contemporary organisations. The three internal mechanisms are ownership concentration, the board of directors and executive compensation. Ownership concentration is the relative amounts of stock owned by individual shareholders and institutional investors. When large-block shareholders (typically own at least 5% of an organisation's issued shares) own a significant percentage of the total shares, they become more active in their demands that the organisation adopt effective governance mechanisms to control managerial decisions. The board of directors are individuals responsible for representing the organisation's owners by monitoring top-level managers' strategic decisions. Board members include insiders (active top-level managers involved in the organisation's day-to-day operations), related outsiders (individuals who are not involved in the organisation's day-to-day operations, but have a relationship with the organisation), and outsiders (individuals who are independent of the organisation). Executive compensation is the use of salary, bonuses and long-term incentives to align managers' interests with shareholders' interests. The external governance mechanism is the market for corporate control, which can manifest in two forms. The first is the purchase of an organisation that is underperforming relative to industry rivals to improve the organisation's strategic competitiveness and earn above-average returns on their investments. This 'hostile takeover', typically, leads to the replacement of ineffective top-level managers. The second is the purchase of an organisation to obtain important resources and expand the business of the acquiring organisation (Hitt et al., 2005).

Corporate governance: acquisition and safety culture integration

In some acquisition scenarios, managers may not be equipped with sufficient communication and change management skills. This is detrimental to the ability of the employees to view and embrace 'the new organisation and its culture' in a positive manner (Kavanagh & Ashkanasy, 2006). Kavanagh and Ashkanasy (2006) suggested the following leadership and change-management strategies to promote smooth transition and reengineering of the culture:

- Appointing a skilled change-management facilitator at the beginning of the acquisition process.
- Using an appropriate approach to change the culture following the acquisition. There are three approaches: immediate (implementing changes within a short period of time), incremental (implementing changes by using negotiations, beginning with those who are keen to embrace the changes), and indifferent (very slow or a lack of changes after the acquisition in which the two entities operate separately). Leaders should be flexible in selecting the approach and may shift from one approach to another depending on the situation.
- Establishing effective communication channels at all levels to inform employees about the stages that should be followed and to outline the outcomes expected from them.
- Selecting those who are willing to embrace the changes before approaching those who resist after allowing enough time for consultation and justification.
- Leading in a positive manner because change is an emotive process. Employees should be changed with dignity by acknowledging contributions and justifying the reasons for change.

Corporate governance: case study

In terms of SSM, safety should be integrated in the organisation's corporate governance mechanisms so that top-level managers consider safety as one of the key factors that influence their strategic decisions. Consider Leighton Holdings (Leighton), a multinational corporation and the largest construction organisation in Australia (Leighton Holdings, 2014) as an example. According to its corporate governance statement, Leighton considers safety performance as one important aspect. Leighton's board of directors is the governance body for safety and they operate through the organisation's Ethics and Compliance Committee. The safety committee members include the three types of board members (insiders, related insiders, and outsiders) as discussed previously. One of the responsibilities of the committee is to monitor and review compliance with applicable legal and regulatory requirements in the areas of safety and make recommendations to the board of directors regarding changes to improve performance. Third-party safety audits are also conducted periodically to review Leighton's level of compliance in its operation. Furthermore, board members attend project site visits and safety briefings, which sometimes occur at remote locations. These activities help the board members to gain an understanding of the opportunities and challenges that can arise within the business and the environments where they operate.

Safety is also a factor that influences the executive compensation at Leighton. For a Leighton executive, 40% of the amount that could be earned as part of the short-term incentive (an annual compensation delivered as a combination of cash and deferred equity which varies up to 100% of fixed remuneration) programme is tailored based on performance against non-financial measures and targets. One of the financial measures is safety, including leadership, sharing of safety learning and total recordable injury frequency rate. On the positive side, Leighton includes the implementation of critical initiatives to improve safety in the non-financial component to ensure that those initiatives are recognised and rewarded. As a penalty measure, on the other hand, a 10% reduction in total short-term incentive is applied when zero fatality is not achieved (Leighton Holdings, 2014).

In order to expand its business, Leighton acquired a 70% shareholding in John Holland in 2000, which then was increased to 100% in 2007. This can be considered as an example of the market for corporate control external governance mechanism. The impact of such an acquisition is that there may be a cultural misalignment (including safety) between the acquiring organisation and the acquired. Corporate governance as well as organisational structure and strategic leadership, discussed in the following sections, are tools to reduce this cultural misalignment at the strategic level.

Organisational structure for safety

Organisational structure specifies the organisation's formal reporting relationships, procedures, controls, and authority and decision-making processes (Hitt et al., 2005). It explains the chain of command, which is the line of authority extending from upper organisational levels to the lowest level, clarifying who reports to whom (Robbins et al., 2012). Strategies and organisational structure are in a reciprocal relationship, although research found that strategy has a stronger influence on structure than the reverse. This means that when the strategies are changed, the organisation should simultaneously consider the structure needed to support the implementation of the new strategies (Hitt et al., 2005). A proper organisational structure, therefore, is also needed to implement safety strategies in the organisation. In a small organisation, this strategy implementation is relatively simple. In a large organisation, on the other hand, organisational policies and structure are complex and should be anchored at the top of the organisation with a control mechanism at each management level. In the context of SSM, the key safety positions should be included and reflected in the organisation structure, for example, a very senior position in the organisation should be appointed for being responsible for safety, and corresponding authorities delegated to this position.

Strategic safety leadership

Strategic leadership is the ability to anticipate, envision, maintain flexibility and empower others to create strategic change as necessary (Hitt et al., 2005). Effective strategy implementation depends on the commitment of all managers to support and lead strategic initiatives within their areas of responsibility (Schermerhorn et al., 2014). Ireland and Hitt (1999) recommended the following six activities to provide effective strategic leadership, which we have adapted to the SSM context:

- *Determining the organisation's vision.* As stated earlier, a vision is an underlying philosophy for the organisation in its operations. Safety should be part of the organisation's overall vision.
- *Exploiting and maintaining core competencies.* Strategic leaders should find ways for safety knowledge to breed still more knowledge to strengthen the organisation's core competencies.
- *Developing human capital.* In the context of SSM, as discussed in Chapter 5 about safety training and learning, people are the most critical resource for an organisation.
- *Sustaining an effective organisational culture.* Strategic leaders should foster a mature and strong safety culture as discussed in Chapter 3.
- *Emphasising ethical practice.* Strategic leaders use honesty, trust and integrity as the foundations of their decisions. Safety, for example, should be implemented not only because of economic reasons, but based on the grounds of morality and human rights, as discussed in Chapter 2.
- *Establishing balanced organisational controls.* Strategic leaders should establish control mechanisms to evaluate the effectiveness of safety strategies. This is discussed further in the next section.

Evaluating safety management strategies

We propose the use of the balanced scorecard introduced by Kaplan and Norton (1996) as a method to evaluate the effectiveness of an organisation's safety management strategies. A balanced scorecard assesses an organisation's performance from four perspectives: financial (growth, profitability and risks), customer satisfaction (satisfaction, loyalty, retention), internal business processes (key processes that create customer and shareholder satisfaction) and learning and growth (a climate that promotes change, innovation and growth) (Hitt et al., 2005). Based on this, we have identified four dimensions to evaluate the effectiveness of safety management strategies, with modified perspectives/dimensions as shown in Figure 8.1:

- The first dimension is financial performance, which can be measured by the organisation's profit, accident compensation costs and safety-related insurance premiums.
- The second dimension is client satisfaction, which can be measured by a satisfaction survey, organisation's reputation and share prices.
- The third dimension is safety culture and climate, which measures the attitudes and perceptions of employees towards safety, employees' safety behaviour and the effectiveness of the safety management system in the organisation (as discussed in Chapter 3).
- The fourth dimension is accident rate, which represents the number of accidents per 100,000 work hours on-site.

Within each dimension, *objectives, measurement indicators, achievement of targets* and *improvement initiatives* should be determined and recorded. It should be noted that beyond these dimensions, there are also other quantitative and qualitative indicators for evaluating construction safety performance. The indicators will depend greatly on the strategy management context in specific organisations.

The balanced scorecard helps managers to link the organisation's long-term safety strategy with its short-term actions. There are four management processes introduced by the balanced scorecard that enable this linking. First, it helps managers translate the safety vision and strategies to an integrated set of objectives and indicators which describe the long-term definition of success. Second, a balanced scorecard lets managers communicate their safety strategy up and down the organisation and link it to departmental and individual objectives. Third, it helps managers undertake and coordinate activities and initiatives that move the organisation towards their long-term safety objectives instead of implementing a wide variety of programmes which may not be relevant to the business. Fourth, it gives the organisation the capacity for strategic safety learning, that is, allowing the organisation to monitor short-term results based on the four dimensions and evaluate as well as modify strategies in the light of recent performance results (Kaplan & Norton, 1996).

Case study

This section presents a SSM case study which was drawn from Lend Lease (LL), a multinational and the second largest construction company in Australia. LL employs over 16,500 employees across the globe, with a revenue of more than $12 billion in 2013 (Lend Lease, 2014). The company is highly committed to safety and strives to operate IIF wherever it has a presence. The strategy context is varied and includes a wide range of market sectors in the construction and property industry. The strategy process consists of three steps: developing, implementing and evaluating IIF strategy. The safety strategy content is to focus on the human side of safety by initiating cultural change so that safety values are embedded into all employees and all stakeholders are involved in and accountable for safety. Data in this case example were collected from the company's website, annual reports, interviews and correspondence with safety managers.

Developing the incident- and injury-free strategy

Although its safety records were much better than the industry average performance, LL recognised that the number of fatalities and serious injuries had reached a plateau despite its advanced system, equipment and processes. LL decided that to achieve a breakthrough, it needs to focus on the human side of safety and to initiate a cultural change whereby every employee is instructed and actively encouraged to put safety first. LL strives to empower its people to believe that they can achieve a workplace free of incidents, injuries and deaths. With this vision in mind, LL launched an IIF safety initiative in 2002, which was a journey to improve safety through the development of a mind-set that is intolerant of any incident and injury (Lend Lease, 2003).

In SSM, safety vision shows the commitment from the top management and is the foundation that upholds the entire process. The following is the safety vision of LL: 'We are committed to operating Incident & Injury Free wherever the Group has a presence. To achieve our Incident & Injury Free vision, it is crucial that we provide our people with the right tools to deliver safe outcomes via our safety management system' (Lend Lease, 2011a). This vision is an underlying philosophy in organisational operations and should be integrated with stakeholder interest. An example of this integration is provided in Figure 8.2.

Under this SSM vision and contents, a specific safety process named ROAD (Risk and Opportunity at Design) was developed and implemented in every construction project. Details of this ROAD process have been discussed in Chapter 6 and also can be found in Zou et al. (2008).

Implementing the incident- and injury-free strategy

LL's IIF strategy is anchored by three objectives and implementation actions: *Owning, Enabling* and *Sustaining*.

Owning – LL believes that the commitment and involvement from all parties at all levels is important for this initiative to succeed. It is essential to create an environment where the workers believe that all injuries are preventable; no injury is acceptable; and schedule, cost or production is not ranked ahead of an injury-free workplace. There are several strategies to support 'Owning'. The first is introducing a series of two-day IIF leadership commitment workshops for managers, designed to realise the potential of IIF. The second is conducting IIF orientation workshops for all employees, while also including them as

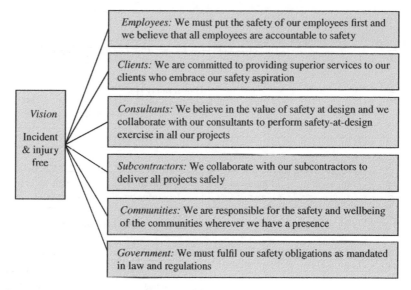

Figure 8.2   Safety vision and stakeholder interests

part of the induction process for new employees. All workshops incorporate the principles of andragogy in their implementation by engaging participants and encouraging them to challenge their perceptions to develop personal commitment. The third is engaging stakeholders to win their commitment to the IIF vision with an aim to make it a core value and a driver for LL in all its operations. Subcontractors are enrolled into IIF workshops while clients are informed about LL's approach to safety (Lend Lease, 2003).

Enabling – The strategies to support this include the organisational alignment with IIF where all the policies, management structure and roles are restructured and redesigned to align with the IIF vision. Lines of accountability and authority were established to identify key positions throughout the company. LL has a Sustainability Committee and one of its main responsibilities is to oversee the organisation's safety function and performance. The committee members are all independent, non-executive directors. Lend Lease's Chief Operating Officer supplies the committee with the information relevant to the committee function, while the Chair of the committee liaises with the Chief Operating Officer on at least a quarterly basis (Lend Lease, 2013a). For implementing safety strategies, Lend Lease has a Group Head of Safety at the top of its organisational structure, along with the Global Safety Leadership Team. Lend Lease operates in four regions: Australia, Asia, Europe and the Americas. A regional head of safety leads safety implementation in each region, while a country safety manager leads safety implementation at the country level. Safety leadership teams are also formed at the regional and country level. The regional operation manager is one of the members at the regional level, while the country manager is a member at the country level. The safety leadership teams meet periodically to ensure that top-level managers are aware of safety performance in their respective business units. Finally, a safety manager implements safety strategies at the project

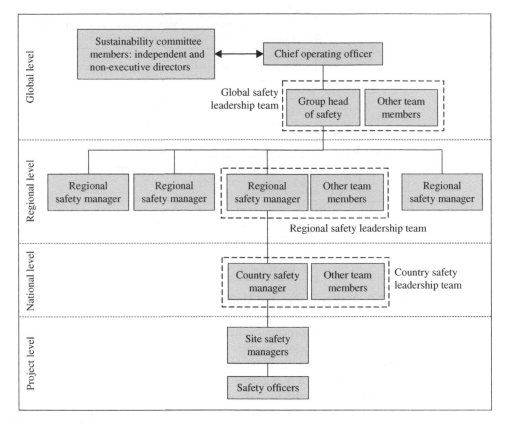

Figure 8.3   Safety management structure

level with the assistance of safety officers who monitor safety performance on the work site. Figure 8.3 illustrates this safety management structure.

Along with the safety-related organisational structure, communication channels are established to ensure that safety information is communicated to all levels of management, aligning the culture and behaviours of the employees to the IIF vision. Safety data collected across each business are reviewed on a monthly basis. They are reviewed first at the project level, then at the business unit and regional level and finally by the Global Safety Leadership Team. This governance system is linked to reward and promotion for positive safety performance and consequences for unsafe behaviours.

LL also maintains a corporate risk register to help employees to identify, assess and mitigate safety risks related to their projects. A process called ROAD was established to consider safety risks early in the design stage (Zou et al., 2008).

In terms of learning, LL has a range of orientation and training programmes to ensure that employees are aware of the health and safety risks associated with their activities and the measures needed to control them. The company developed a global online Safety Passport course with training modules on IIF and the Global Minimum Requirements (GMRs) that set out the minimum environment, health and safety standards for controlling the risks associated with LL operations. Besides the Safety Passport training, employees and sub-contractors must also undertake technical and management training to enable them to deal with the specific health and safety risks in their roles. Contractors who perform specialised/high-risk operations are required to produce proof of competence before starting work (Lend Lease, 2011b).

Another programme, called Springboard, is designed to help employees realise and reach their potential in their work and personal life. Springboard brings employees from different cultures and business units together for an intensive four-day personal development experience. With its community partners, Springboard also provides opportunities for employees to contribute and 'give back' to the community where LL operates (Lend Lease, 2011d).

Sustaining – LL aspires to sustain and lead the industry by sharing the benefits of the organisational transformation with their stakeholders. LL invests in research, innovation and benchmarking to continually redefine the vision. It also sustains leadership commitment by reviewing, recognising and rewarding behaviour of leaders for achieving the IIF vision. A 'living' communication plan is used to capture feedback from all stakeholders. The IIF system is also evaluated periodically to facilitate the transformation process towards the IIF vision (Zou et al., 2006).

LL also recognises the necessity of continually maintaining and developing its safety culture. LL regularly engages external workplace safety consultants to provide independent advice to further embed safety culture into business and work practices and to audit its performance (Lend Lease, 2005). Starting in 2009, LL trained its managers and senior staff in the Uncompromising Leadership programme to develop a safety-focused mindset. LL realises that safety culture starts at the top and that there is a need for senior management to visibly demonstrate their commitment to safety by actively engaging stakeholders and integrating safety into business practices (Lend Lease, 2011c).

Research and development is another platform of LL to sustain its performance. In 2006, using safety-in-design principles, LL released a Falls Mandate which aims to eliminate risks related to falls. The Falls Mandate operates through an authorisation hierarchy to ensure senior management engagement in the planning process and the consistent application of standards. This has led to greater awareness towards fall-related risks and transparency in reporting, demonstrated by a significant increase in the reporting of falls of materials, while at the same time the number of people falling has decreased (Lend Lease, 2006).

In 2008, LL adopted an online safety reporting system called the Safety Dashboard. Developed by the LL IT team, the Safety Dashboard provides a central reference point for all key construction safety data from all LL projects across the world (Lend Lease, 2009). In 2010, embracing the necessity of safety research–practice nexus to improve performance, LL collaborated with researchers to investigate the economic benefits of safety (Sun & Zou, 2010; Zou et al., 2010).

Evaluating the incident- and injury-free strategy

LL's safety management system and GMRs set specific requirements for performance monitoring and evaluation. The results of checks, inspections and audits are recorded in an online reporting tool, called WebCare, and the data used to identify problem areas and implement actions to deliver improvements. LL records loss time incidents across its business units. Incidents which have a clear potential to result in serious injury are reported via WebCare to allow LL's Safety Leadership Teams to better understand the circumstances, communicate lessons learnt and manage risks proactively. Serious incidents are thoroughly investigated while Root Causes Analysis (RCA) is conducted to identify the underlying causes of incidents and the need to change (Lend Lease, 2011b). In 2013, LL achieved zero fatalities. This is supported by 6% and 9% growth in revenue and profit respectively (Lend Lease, 2013b). Concurrently, in partnership with a university, LL conducted a study to measure the benefit of their investments in safety and the results show that due to superior safety performance, their investment in IIF safety strategy has generated a positive return in all their case projects.

LL tracks employee perceptions and benchmarks people management performance against best practice in the industry and against global high-performing organisations, using a global employee engagement survey (Lend Lease, 2011d). This is similar to the safety climate survey that we propose as one of the dimensions of the balanced scorecard method for evaluating strategy effectiveness.

It should be pointed out that although it is important to evaluate the economic return of safety strategy development and implementation from cost–benefit perspectives, it is equally important to be aware of the imperative of corporate social responsibility, where the corporation must fulfil its duty of care to its workforce and to everyone with whom it has business dealings, in addition to the general public. Protecting human life, safety and health, in many situations cannot be simply measured in monetary terms but must be viewed from a moral and ethical standpoint, as discussed in Chapter 2.

Conclusions

This chapter has provided an overview of SSM, including its development, implementation and evaluation, from fundamental theories and principles to a case study. This chapter has integrated strategic factors discussed in the previous chapters into an SSM framework. First, construction organisations and stakeholders should understand the economic benefits of safety so that they are willing to invest in and support safety implementation. Second, safety culture should be developed, not only within an organisation, but also across the supply chain. Third, safety at design, including architectural and engineering designs, is a practical way to improve safety by mitigating safety risk early during the project lifecycle and by improving communication between designers and contractors, use of risk management and BIM techniques.

Fourth, all employees should be equipped with the necessary safety skills and knowledge to enable them to be safety leaders in their workplace. Fifth, the safety learning process should include the principles of andragogy in order to be effective, particularly for providing safety training to workers who may already have a certain level of working experience. Sixth, safety research should consider its relevance, implication and nexus relationship to practice. Seventh, a balanced scorecard method can be used to evaluate the effectiveness of SSM of construction organisations. A case study has been presented to show how all these factors can be applied in practice.

In conclusion, we argue that SSM is a feasible way to achieve the desired maturity of safety management and its integration into decision-making processes in the construction industry. Commitment from and collaboration among key stakeholders (clients, contractors, subcontractors, consultants and governments) are crucial for such an integration to become a reality. It should be noted, however, that implementing strategies is never a simple task, thus two key barriers are worth mentioning. First, there may be misalignment from the company's boardroom decision-making to its implementation throughout the organisation (Sunindijo & Zou, 2013). Hrebiniak (2006) explained that there may be separation between strategy planning and strategy execution, where the planners (the 'smart' people) develop plans that the 'grunts' (people not quite as 'smart') have to make the plans work. When things go awry, the problem is attributed to the 'grunts'. The second barrier is resistance to change. People, who have been in the industry for many years, with a certain level of working experience, may believe that they know everything and know how to do their work and do not like to change. They tend to resist new things that they do not understand or do not feel comfortable with. Although some may realise that there is a better and safer way to work, as they have been working in a certain way for a long time, it could be difficult for them to change (Sunindijo & Zou, 2013).

To this end, it is important to point out that SSM development, implementation and evaluation are an iterative process where constant SSM monitoring, SSM review and consultation with stakeholders should be undertaken and feedback provided to the boardroom and senior management so that modification of the SSM contents, process and context are modified and improved to better suit the frontline and on-site workplace. It is also worth pointing out once more that top management commitment to constantly championing safety, and the bottom-up involvement and cooperation from workforce are vitally important for any SSM to be successful.

References

ASX Corporate Governance Council. (2007). *Corporate Governance Principles and Recommendations with 2010 Amendments.* Sydney: ASX Corporate Governance Council.

Fang, D., & Wu, H. (2012). Safety culture in construction projects. Paper presented at the CIB W099 International Conference on Modelling and Building Health and Safety, Singapore.

Fluor Corporation. (2014). Fluor's Sustainable Commitment to Health, Safety, and the Environment Retrieved 15 Jan, 2014, from http://www.fluor.com/sustainability/health_safety_environmental/Pages/default.aspx

Hale, A., & Hovden, J. (1998). Management and culture: The third age of safety. In A. Feyer & A. M. Williamson (Eds.), *Occupational Injury: Risk Prevention and Intervention*. London: Taylor and Francis.

Hamel, G. (2000). *Leading the Revolution*. Boston: Harvard Business School Press.

Hitt, M. A., Ireland, R. D., & Hoskisson, R. E. (2005). *Strategic Management: Competitiveness and Globalization* (6th ed.). Mason: South-Western.

Hopkins, W. E., cHopkins, S. A., & Mallette, P. (2005). *Aligning Organizational Subcultures for Competitive Advantage: A Strategic Change Approach*. New York: Basic Books.

Hrebiniak, L. G. (2006). Obstacles to effective strategy implementation. *Organizational Dynamics, 35*(1), 12–31.

Hudson, P. (2007). Implementing a safety culture in a major multi-national. *Safety Science, 45*(6), 697–722. doi: 10.1016/j.ssci.2007.04.005

Ireland, R. D., & Hitt, M. A. (1999). Achieving and maintaining strategic competitiveness in the 21st century: The role of strategic leadership. *Academy of Management Executive, 13*(1), 43–57.

Johnson, G., Scholes, K., & Whittington, R. (2008). *Exploring Corporate Strategy* (8th ed.). Harlow, England: Prentice Hall.

Kaplan, R. S., & Norton, D. P. (1996). Using the balanced scorecard as a strategic management system. *Harvard Business Review*, Jan–Feb, 178–192.

Kavanagh, M. H., & Ashkanasy, N. M. (2006). The impact of leadership and change management strategy on organizational culture and individual acceptance of change during a merger. *British Journal of Management, 17*(S1), S81–S103.

King, M. L., Jr. (1963). I Have a Dream Retrieved 25 August, 2014, from http://www.americanrhetoric.com/speeches/mlkihaveadream.htm

Kirkpatrick, D. L. (1979). Techniques for evaluating training programs. *Training and Development Journal*, June, 178–192.

Kirkpatrick, D. L. (1996). Great ideas revisited: Revisiting Kirkpatrick's four-level model. *Training & Development, 50*(1), 54–57.

Leighton Holdings. (2014). *Leighton Holdings Limited Annual Report 2013*. Sydney: Leighton Holdings.

Lend Lease. (2003). *Strength through Focus: 2003 Annual Report to Shareholders*. Sydney: Lend Lease.

Lend Lease. (2005). *2005 Annual Report to Shareholders*. Sydney: Lend Lease.

Lend Lease. (2006). *2006 Annual Report to Shareholders*. Sydney: Lend Lease.

Lend Lease. (2009). *Creating Sustainable Landscapes: 2009 Annual Report to Shareholders*. Sydney: Lend Lease.

Lend Lease. (2011a). Corporate Governance Retrieved 9 August, 2013, from http://www.lendlease.com/en/worldwide/about-us/corporate-governance.aspx

Lend Lease. (2011b). Health and Safety: Detail Retrieved 14 July, 2014, from http://www.lendleasesustainability.com/sustainability/healthSafetyDetail.html#/health-and-safety-detail

Lend Lease. (2011c). *Lend Lease Securityholder Review 2011*. Sydney: Lend Lease.

Lend Lease. (2011d). Our People: Detail Retrieved 14 July, 2014, from http://www.lendleasesustainability.com/sustainability/ourPeopleDetail.html#/our-people-detail

Lend Lease. (2013a). *Lend Lease Annual Report 2013*. Sydney: Lend Lease.

Lend Lease. (2013b). *Lend Lease Securityholder Review 2013*. Millers Point: Lend Lease.

Lend Lease. (2014). About Lend Lease Retrieved 18 Feb, 2014, from http://lendlease.com/

Lingard, H., & Rowlinson, S. (2005). *Occupational Health and Safety in Construction Project Management*. Oxon: Spon Press.

McGeorge, D., & Zou, P. X. W. (2013). *Construction Management: New Directions* (3rd ed.). Chichester, UK: Wiley.

Pillay, M. (2014). Taking stock of zero harm: A review of contemporary health and safety management in construction. Paper presented at the CIB W099, Lund, Sweden.

Price, A. D. F., & Newson, E. (2003). Strategic management: Consideration of paradoxes, processes, and associated concepts as applied to construction. *Journal of Management in Engineering, 19*(4), 183–192.

Robbins, S. P., Bergman, R., Stagg, I., & Coulter, M. (2012). *Management* (6th ed.). Australia: Pearson.

Schermerhorn, J. R., Davidson, P., Poole, D., Woods, P., Simon, A., & McBarron, E. (2014). *Management* (5th Asia-Pacific ed.) Milton, Australia: John Wiley & Sons Australia.

Sun, A. C. S., & Zou, P. X. W. (2010). Understanding the true costs of construction accidents. Paper presented at the CIB World Congress, Salford Manchester UK.

Sunindijo, R. Y., & Zou, P. X. W. (2013). Aligning safety policy development, learning and implementation: From boardroom to site. Paper presented at the CIB World Building Congress 2013, Brisbane, Australia.

de Wit, B., & Meyer, R. (2005). *Strategy Synthesis: Resolving Strategy Paradoxes to Create Competitive Advantage (Text and Readings)* (2nd ed.). London, UK: Thomson Learning.

Zou, P. X. W., & Sunindijo, R. Y. (2013). Skills for managing construction safety risks, implementing safety tasks and developing positive safety climate. *Automation in Construction, 34*, 92–100.

Zou, P. X. W., Windon, S., & Mahmud, S. H. (2006). Culture change towards construction safety risks, incidences and injuries: Literature review and case study. Paper presented at the CIB W099 International Best Practices, 28–30 June, Beijing, China.

Zou, P. X. W., Redman, S., & Windon, S. (2008). Case studies on risk and opportunity at design stage of building projects in Australia: Focus on safety. *Architectural Engineering and Design Management, 4*(3–4), 221–238.

Zou, P. X. W., Sun, A. C. S., Long, B., & Marix-Evans, P. (2010). Return on investment of safety risk management system in construction. Paper presented at the CIB World Congress, 11–13 May, Salford Manchester, UK.

Zou, P. X. W., Marsh, D., Evans, M., Sherrard, A., & Howard, J. (2014a). Changing construction safety culture and improving safety outcome by design thinking and co-production: Research proposal and preliminary results. Paper presented at the CIB W099 Achieving Sustainable Construction Health and Safety, Lund, Sweden.

Zou, P. X. W., Sunindijo, R. Y., & Dainty, A. R. J. (2014b). A mixed methods research design for bridging the gap between research and practice in construction safety. *Safety Science, 70*, 316–326.

Bibliography

This bibliography provides an overview of the currently available books relevant to construction safety. It also shows the gap which this book attempts to fill in.

Goetsch, D. L. (2012) *Construction Safety & Health*, (2nd ed.), Prentice Hall, Upper Saddle River, New Jersey, USA.

> This book offers various principles of construction safety and health including accident causation theories, Occupational Safety and Health Administration (OSHA) construction standard and related safety practices, safety program implementation, and construction health issues. This book particularly focuses on the requirements stated by OSHA and other regulators. This book also provides recommendations, activities, and systems that can be applied to manage and comply with safety and health regulations.

Griffith, A. (2011) *Integrated Management Systems for Construction: Quality, Environment and Safety*, Harlow: Prentice Hall.

> This book covers relevant international standards for quality, safety, and environment (ISO9000, ISO14001, and ISO18001) and discusses approaches for integrating them.

Griffith, A. and Howarth, T. (2001) *Construction Health and Safety Management*, Edinburgh: Pearson Education.

> The main aim of this book is to help contractors implement health and safety management system in an appropriate way within the scope of the Construction (Design and Management) Regulations 1994 and other existing legislations in the UK prior to the year 2000. The book discusses the topics that should be focused upon to implement health and safety management effectively: management structure, human factors, health and safety policy development, roles and responsibility of safety personnel, risk assessment, construction health and safety planning, and system auditing.

Hill, D. C. (Editor) (2014) *Construction Safety Management and Engineering*, (2nd ed.). ASSE, Des Plaines, IL, US.

> This edited book consists of five parts: before the work commences; components of safety implementation process; liability and regulations, instructions on managing specific job hazards; and other considerations including ergonomics, construction safety in healthcare facilities, and communication during a crisis.

Holt, A. S. J. (2005) *Principles of Construction Safety*, Oxford: Blackwell.

> This book explains the principles of construction safety and offers practical guidance on best practice to implement safety in the construction context. It covers a wide range

Strategic Safety Management in Construction and Engineering, First Edition.
Patrick X.W. Zou and Riza Yosia Sunindijo.
© 2015 John Wiley & Sons, Ltd. Published 2015 by John Wiley & Sons, Ltd.

of topics including safety performance measurement, safety policy development, risk assessment, safety training, understanding people, legal requirements, and safe work method statements.

Hopkins, A. (2012) *Disastrous Decisions: The Human and Organisational Causes of the Gulf of Mexico Blowout*, North Ryde: CCH Australia.

This book provides an in-depth discussion concerning the human and organisational factors that contributed to the Deepwater Horizon disaster in 2010 (oil drilling rig explosion) resulted in the loss of 11 lives. This book particularly focuses on the importance of human factors in safety.

Howarth, T. and Watson, P. (2009) *Construction Safety Management*, Chichester: Wiley-Blackwell.

Written in the context of the UK construction industry, this book serves to inform readers about health and safety knowledge and its practice. It addresses safety legislations in UK and their requirements as well as providing guideline to manage construction health and safety system. It also considers the concept of health safety culture and recommends practical initiatives and tools for promoting a positive safety culture.

Hughes, P. and Ferrett, E. (2011) *Introduction to Health and Safety in Construction*, (4th ed.). Oxford, UK: Butterworth-Heinemann.

This book covers a wide range of health and safety aspects in construction, particularly for those who are interested to get safety and health certification from the UK-based National Examination Board in Occupational Safety and Health (NEBOSH).

Levitt, R. E. and Samelson, M. (1993) *Construction Safety Management*, (2nd ed.). New York: John Wiley & Sons.

This book drew its contents from the Stanford Construction Safety Research Program started in 1970. It consists of seven parts: the benefits of safety; the role of the Chief Executive, particularly to develop safety culture; the role of the job-site manager; the role of the foreman; the role of the safety professional; considering safety in procuring contractors; and other potential management tools and technologies to manage construction safety.

Li, R. Y. M. and Poon, S. W. (2013) *Construction Safety*, Heidelberg: Springer.

This book includes a range of recent research studies on construction safety, including the effectiveness of different management tools to reduce accident rates, the view of senior and junior construction personnel on safety, using information technology to support safety knowledge management, case studies on safety measures implementation, and discussion on compensation related to construction accidents.

Lingard, H. and Rowlinson, S. (2005) *Occupational Health and Safety in Construction Project Management*, Oxon, UK: Spon Press.

This book addresses various aspects of occupational health and safety (OHS) including historical development, legal issues, organisational issues, social and psychological perspectives, and on-site conditions. It promotes a multidisciplinary approach to safety risks to ensure that OHS is managed in a holistic manner.

McAleenan, C. and Oloke, D. (Editors) (2010) *ICE Manual of Health and Safety in Construction*, London: Thomas Telford.

This edited book explains the issues of managing OHS throughout project life cycle and recommends methods to control health and safety hazards.

Kelloway, E. K. and Cooper, C. L. (Editors) (2011) *Occupational Health and Safety for Small and Medium Sized Enterprises*, Cheltenham, UK: Edward Elgar.

The book discusses different topics relevant to small and medium organisations in the construction industry including challenges and potential solutions, practical techniques to manage safety, workplace violence, stress management, sexual harassment and health and well-being.

Poon, S. W., Tang, S. L., and Wong, F. K. W. (2008) *Management and Economics of Construction Safety in Hong Kong*, Hong Kong University Press, Hong Kong.

Using the context of the Hong Kong construction industry, this book provides information and analysis on various aspects of construction safety management and economics, including safety programs and their performance, safety management systems, safety legislation, safety auditing, accident investigation, site supervision and financial costs, social costs, and human pain and suffering costs of construction accidents.

Reason, J. (2008) *The Human Contribution*, Surrey, England: Ashgate.

This book discusses how a human can be considered as a hazard or a hero. In this book, the author gave a number of examples to demonstrate these two seemingly contrasting perspectives in action.

Roughton, J. E. and Mercurio, J. J. (2002) *Developing an Effective Safety Culture: A Leadership Approach*, Woburn, MA: Butterworth-Heinemann.

This book offers a comprehensive approach for developing safety culture through building safety management systems and ultimately developing a safety program that supports safety culture.

Index

Strategic Safety Management in Construction and Engineering, First Edition.
Patrick X.W. Zou and Riza Yosia Sunindijo.
© 2015 John Wiley & Sons, Ltd. Published 2015 by John Wiley & Sons, Ltd.

Printed and bound by CPI Group (UK) Ltd, Croydon, CR0 4YY

07/02/2024

08233445-0003